S0-ATV-984

An Introduction to ECOLOGY AND POPULATION BIOLOGY

The Author

Thomas C. Emmel is assistant professor of zoology at the University of Florida. Since 1968 he has lectured to some two thousand students annually in the course from which this book developed. In 1971–72 he received the Outstanding Professor Award for his undergraduate teaching in biology. He is a graduate of Reed College, received his doctorate in population biology from Stanford University, and was a postdoctoral fellow in genetics at the University of Texas. His field research in ecology and population biology has included numerous expeditions to Mexico and the Caribbean and to both Central and South America.

Nature is but a name for an effect,
Whose cause is God.

William Cowper

In Nature's infinite book of secrecy
A little I can read.

William Shakespeare

An Introduction to ECOLOGY AND POPULATION BIOLOGY

Thomas C. Emmel
UNIVERSITY OF FLORIDA

W · W · NORTON & COMPANY · INC ·
NEW YORK

First Edition

Library of Congress Cataloging in Publication Data

Emmel, Thomas C
An introduction to ecology and population biology.

Includes bibliographical references.
1. Ecology. 2. Population. 3. Man—Influence on nature. I. Title.
QH541.E45 1973 574.5'24 72-14170
ISBN 0-393-06393-3
ISBN 0-393-09371-9 (pbk.)

Published simultaneously in Canada
by George J. McLeod Limited, Toronto
Illustration credits appear on page viii.
This book is printed on recycled paper.

PRINTED IN THE UNITED STATES OF AMERICA

1 2 3 4 5 6 7 8 9 0

CONTENTS

Illustration Acknowledgements

All uncredited photographs are by the author.

FRONTISPIECE: NASA. FIG. 1-2: Redrawn from Heinrich Walter, *Die Vegetation der Erde* (Jena: Veb Gustav Fischer-Verlag, 1964) after Beard in T. W. Richards, *The Tropical Rain Forests,* Cambridge, 1952. FIG. 1-3: Redrawn from Elna Bakker, *An Island Called California* (Berkeley and Los Angeles: University of California Press, 1971). FIG. 1-4: Redrawn from Orlando Park, *Ecology* 19 (1938): 219. Copyright by The Ecological Society of America. FIG. 2-1: Redrawn from Frank B. Salisbury, *Science* 136 (1962): 17–26. Copyright 1962 by the American Association for the Advancement of Science. FIG. 2-2: After Fig. 1, "The carbon cycle in the ecosystem" from *Environmental Insight* by Robert M. Chute (Harper & Row, 1971), p. 45. FIG. 2-9: A, B, D: Charles L. Hogue. E: Redrawn from Elna Bakker, *An Island Called California* (Berkeley and Los Angeles: University of California Press, 1971). FIG. 2-10: Redrawn from Elna Bakker, *An Island Called California* (Berkeley and Los Angeles: University of California Press, 1971.) FIG. 2-12: Data from H. T. Odum. FIG. 2-13: Redrawn from Arthur S. Boughey, *Ecology of Populations,* © 1968, The Macmillan Company, New York. FIG. 3-1: Redrawn from Francis E. Lloyd, *The Carnivorous Plants,* © 1942, The Ronald Press Company, New York. FIG. 3-2: Original painting by Anthony Angell. FIG. 3-3: Original painting by Byron B. Wolfe. FIGS. 3-4, 3-5: From R. W. Ficken et al., *Science* 173 (1971): 936–939. Copyright 1971 by the American Association for the Advancement of Science. FIG. 3-6: A, B: Charles L. Hogue. FIG. 3-9: A: Julian P. Donahue. B, C: Charles L. Hogue. FIG. 3-10: B: Julian P. Donahue. FIG. 3-11: Julian P. Donahue. FIG. 3-12: Redrawn by permission of the publishers from *Life: An Introduction to Biology* by George Gaylord Simpson, Colin S. Pittendrigh, Lewis H. Tiffany, © 1957 by Harcourt Brace Jovanovich, Inc. FIG. 3-13: After Gause, 1934. FIG. 3-14: Redrawn from Robert MacArthur and Joseph Connell, *The Biology of Populations* (New York: John Wiley & Sons. Inc., 1966). FIG. 4-3: Redrawn by permission of the author from Irven DeVore, *Primate Behavior* (New York: Holt, Rinehart and Winston, Inc., 1965). FIG. 4-4: Redrawn from Archie F. Carr, "Orientation Problems in the High Seas Travel and Terrestrial Movements of Marine Turtles," *American Scientist* 50 (1962). FIG. 4-5: From Carr and Hirth, *American Museum Novitates* no. 2091 (1962). FIG. 4-6: Redrawn from U.S. Fish and Wildlife Service material. FIG. 5-3: Redrawn from H. G. Andrewartha and L. C. Birch, *The Distribution and Abundance of Animals.* Copyright © 1954 by the University of Chicago Press. FIG. 5-5: Lynx and hare redrawn from the book *Palmer's Fieldbook of Mammals* by E. Laurence Palmer. Copyright © 1957 by E. Laurence Palmer. Published by E. P. Dutton & Co., Inc., and used with their permission. FIG. 5-6: Fox and lemming redrawn from the book *Palmer's Fieldbook of Mammals* by E. Laurence Palmer. Copyright © 1957 by E. Laurence Palmer. Published by E. P. Dutton & Co., Inc., and used with their permission. Graph after Chitty, 1950. FIG. 6-2: After Fig. 9, "A diagram of forest succession from annual weeds to mature forest" in *The Ecology of Man: An Ecosystem Approach* by Robert Leo Smith (Harper & Row, 1972, p. 15). FIG. 6-3: Redrawn from Elna Bakker, *An Island Called California* (Berkeley and Los Angeles: University of California Press, 1971). FIG. 6-8: Charles L. Hogue. FIG. 6-10: Charles L. Hogue. FIG. 6-11: Charles L. Hogue. FIG. 7-3: Gordon H. Orians. FIG. 7-4: Charles L. Hogue. FIG. 7-5: James Rose. FIG. 7-6: David Ehrenfeld. FIG. 8-1: Chuck Woods, IFAS, University of Florida. FIG. 8-2: Top: Los Angeles County Air Pollution Control District Photo. FIG. 8-3: Rondal Partridge. FIG. 8-4: Chuck Woods, IFAS, University of Florida.

PREFACE

Environmental and population crises have awakened widespread public interest in *ecology*, a word that prior to the 1960s seemed to refer solely to a little-known backwater area of biology. Now thousands of people are daily involved in making ecological decisions, from buying phosphate-free detergents to joining conservation organizations to turn their concern for the environment into political action. Yet, because this public interest in ecology has developed so recently, few of us have had the opportunity to obtain a broadly informative introduction to the subject.

This book was written for the educated general public and for students who want a good basic understanding of their environment, the need for preserving it, and the value of the natural populations that depend on it. The first half of the book develops an understanding of basic ecological principles and how they are involved in natural populations of plants and animals. After this preparation the remaining chapters apply these principles to problems of the human population, including population growth and the wide spectrum of environmental pollution and landscape alteration.

This book came to be written purely and simply because no concise elementary presentation of ecology was available. In the comprehensive biological sciences courses taken by several thousand undergraduates yearly at the University of Florida, a primary problem has been to find a text that would emphasize the ecological approach without becoming needlessly detailed, would treat the salient aspects of population biology without bogging down in theoretical models, would emphasize man, and would be written in a style and depth appropriate to a course enrolling both science and non-science majors. While many good books of collected readings were available, a text of appropriate length built around the theme of ecology and population biology seemed hard to find. Thus this book emerged from both educational need and practical experience in teaching ecology to students with no previous college course in biology. But it is intended equally for all concerned readers who want a better biological understanding of the need for environmental conservation and population control.

I am indebted to many colleagues, students, and friends who have assisted in the evolution and writing of this book. I particularly wish to thank Michael M. Sligh of the University of Florida for his constant encouragement and careful editorial assistance on the manuscript. Three fine secretaries, Eugenia Kelly, Mary H. Sligh, and Ruth F. Smith, have worked long and diligently in typing the manuscript. In preparing this book, I have also appreciated the careful assistance and patience of Kenneth B. Demaree, Mary Pell, and the editorial staff of W. W. Norton & Company. To all, I am most grateful.

THOMAS C. EMMEL

Gainesville, Florida
November 1972

An Introduction to ECOLOGY AND POPULATION BIOLOGY

The planet earth is a spaceship with limited supplies and a defined ecology. It is the only body in our solar system capable of supporting life as we know it. Will we render it uninhabitable, or learn to take care of the ecology of our ship in time?

1 POPULATIONS AND ECOLOGY

Why Study Ecology and Population Biology?

Ecology as a word and as a concern has come into its own in the 1970s. The average person today has heard more about "ecology" than any other area in biology. Daily he is inundated with the flow of newspaper and periodical articles, news broadcasts, and advertisements, all telling him the latest on the ecological crises or potential environmental catastrophes facing his community or even his world. Atomic blasts under a remote Aleutian island, oil spills off a coastal beach, mercury in the consumer's swordfish, DDT on his lettuce, strontium-90 in his milk, a sea-level canal in Panama that may allow poisonous sea snakes to move from the Pacific Ocean into pleasant Caribbean tourist beaches—all clamor for the intelligent citizen's attention, evaluation, and action where possible. Yet ecology is probably the most complex and least understood area of biology today, while being the biological discipline most important and relevant to the future of our world. Thus one purpose of this book is to provide an overall, rigorous, yet easily understandable introduction to ecological concepts and the flow of life, energy, and materials in the natural world.

Since the environmental dangers to the well-being of the human species are in large part the result of population increase, we must be especially concerned with the ways in which populations of organisms, particularly man, grow and function. In particular, how are they limited or left unregulated by natural as well as man-caused factors? Thus a second goal of this book is to familiarize the reader with the basic principles involved in the biology of natural populations of plants and animals, and then to apply these lessons and analyses to human population problems, including population growth and the broad spectrum of environmental pollution and landscape alteration. Familiarity with these points should give the reader a clearer view of the importance and immediacy of the population problems and ecological decisions facing us, and we may hope that your generation will provide the wise conclusions and actions necessary to preserve and improve human life and society as we have come to know it.

With these goals in mind, let us now look into the basic organization of ecology as a science and the biological systems it encompasses.

The Levels of Organization of Ecology

Ecology is the study of the interrelationships between organisms and their environment. It thus deals with nearly all levels of organization of life on earth, from the individual animal or plant to the whole community of organisms living in an area to the effects on these organisms of the climatic and even geologic factors that make up their physical surroundings. In the ultimate and broadest sense, ecologists (students of ecology) are involved in "building an understanding of the role of living things in the structure and function of the universe." * In practice, ecologists tend to specialize in one

* Robert M. Chute, *Environmental Insight* (New York: Harper & Row, 1971), p. 43.

particular level of organization in their research: the individual, the population, the community, or the ecosystem.

Autecology is the study of the individual organism. This is the smallest working unit for an ecologist, and the investigator is interested in what the individual requires and tolerates, its way of life, its functioning in the environment throughout all stages of its life cycle.

Population ecology is the study of populations of organisms. A population consists of all the individuals of a species living in an area. Population ecologists wish to explain the behavior of populations—their stability, rapid increase, or decline. In the southeastern United States, water hyacinths are far too numerous on the rivers and lakes, causing navigational and recreational problems for man, while Ivory-bill Woodpeckers are on the verge of extinction. Cockroaches live by the millions in old student housing on many university campuses, while the wilderness-loving Whooping Cranes number less than forty individuals. People in the United States expend great amounts of money to kill hyacinths and cockroaches and save Ivory-bills and Whooping Cranes. The reasons for these opposing situations of explosion and extinction are not only interesting from the practical and aesthetic viewpoints, respectively. They teach lessons that may be applied to man's current population explosion (man was a species that exhibited near stability in population size until about 1600 A.D).

Community ecology is the study of biotic communities. A biotic community is composed of all the organisms of all species living in a particular area. A pond community, for instance, with its plant and animal inhabitants, may contain a very complex set of biological interrelations which, when studied in detail, provide invaluable information on the flow of energy and chemical elements through organisms and eventually to man.

An **ecosystem** includes both the biotic community and the physical environment in an area. The Mojave Desert in southern California is a fairly simple terrestrial ecosystem, with rather few plant and animal species, high insolation (amount of sunlight), and a limited amount of rainfall (normally less than ten inches) each year. On the other hand, a

Fig. 1-1. The Mojave Desert in southern California (top), a relatively simple and orderly ecosystem with few species of organisms, and a tropical rain forest, a place of great diversity yet intricate orderliness in an evolutionary and ecological sense.

tropical rain forest, which receives some 150 to 200 inches of rain per year, is the most complex terrestrial ecosystem in existence, with thousands of plant and animal species found in a single square mile (Fig. 1-1). Both the desert and the tropical rain forest ecosystems will remain in the same general condition over hundreds of years if undisturbed by natural disaster, man, or climatic changes. A home garden or an orange grove, though, is a very artificial ecosystem that is highly dependent on man for its continued existence. Why do these man-created ecosystems differ so greatly in stability from natural ecosystems? Why are gardens and orange groves more desirable to man? Why won't they persist if man does not continue to maintain them? We will look at some of the answers to such questions in the remainder of this chapter and Chapter 2.

Evolution and Orderliness in Ecological Systems

An ecosystem's component organisms and the factors making up the physical environment are organized into a system that is analogous in many ways to the organization of an individual organism. The interrelation between organs in the body, or components of an ecosystem, is not haphazard. It has a definite history of development, a particular spatial orientation, an involvement of time in operation of the system, and an involvement of specific energetic sequences (food is consumed, processed, and the molecules re-utilized). Thus ecosystems are characterized by four major kinds of orderliness: **evolutionary, spatial, temporal,** and **metabolic** orderliness.

Evolutionary Orderliness

The species of plants and animals living in a given region are the most recent products of organic evolution. Each is adapted or adjusted to the particular environment it lives in through **differential reproduction,** which results in what we call **evolution,** that is, change in a species' characteristics, even the development of new species. Differential reproduction simply means that the best adapted individuals of a species tend to leave more offspring on the average than do less well adapted genetic variants. We call this result "natural selection" or "survival of the fittest individuals." When we consider the whole biotic community in an ecosystem, we see that the plant and animal species have not evolved independently of each other. Instead of a miscellaneous collection of kinds of organisms, the species composition of an ecosystem is *ordered.* There has been a coevolution of the plant and animal components of an ecosystem; mutual adjustments of species have taken place through natural selection. Thus the biotic parts of the ecosystem form a cohesive whole because they have shared a common recent evolutionary history, adjusting to each other and to the special set of environmental factors prevailing in that terrestrial or aquatic area.

Spatial Orderliness

An ecosystem does not contain a miscellaneous assortment of adaptively unrelated species; nor are these species distributed in a casual manner. The placement of individuals is specific for different species. Each animal or plant lives in a particular place, which may be in soil, above soil, in the tops of trees, in flower heads, in the shallow margin of a pond, or under rocks. This spatial arrangement is determined by each species' **ecological niche,** which is simply a way of life unique to each species. The niche includes the species' physical **habitat** (the place where the organisms live) and **adaptive strategy** (how the species acquires energy and makes its living). The problem of defining the concept of niche has been a subject of much debate among ecologists, but almost all agree that it is more than just a place on a shelf, that is, a physical space inhabited by the organism. It involves the organism's activities as well.

Spatial orderliness is frequently manifested as **stratification** in an ecosystem; the living organisms in an aquatic or terrestrial community seem to be arranged in vertical layers. In a deciduous forest, for example, one passes through wholly different sets of species as one goes from the soil up through the lower-story vegetation to the uppermost forest canopy. The extent of this stratification becomes quite complex in tropical rain forests (Fig. 1-2), where the number of species is greatest. Even animal communities often exhibit stratification in their physical distribution among foraging areas or nest sites (Fig. 1-3).

Temporal Orderliness

Organisms in an ecosystem are not randomly active through the 24-hour day; instead, each species has a particular time of maximum activity. This often permits many more species to exist in an area than if all species carried on their activities at one particular time or if their activity periods commonly overlapped. In **daily periodism** an organism's principal activities, such as feeding and moving about, are limited to certain hours in the 24-hour day; some activities may be at night and others in the daytime. A particular type of daily periodism is **nocturnal** or **diurnal periodism,** in which the activities of the organism take place only at night or only in the daylight hours (Fig. 1-4). On a longer time scale, **lunar periodism** is characteristic of many marine communities especially. Microscopic animals and plants called plankton will migrate vertically in the ocean depths in accord with the phase of the moon. **Breeding** or **reproductive seasons** are common examples of temporal orderliness in biotic communities: species utilize particular times of the year to reproduce (in the case of animals, usually when the maximum amount of food is available for rearing young).

Metabolic Orderliness

In an ecosystem, transformations of energy and materials follow definite, orderly paths. Materials are cycled in **biogeochemical cycles.** This term simply refers to the cycling of chemical elements through the organisms in a biotic (living) community and their physical environment. Few chemical elements are permanently lost from the community; hence transformations of matter are said to be cyclical. Energy is

Fig. 1-2. Stratification in a tropical rain forest, as seen in vertical transect.

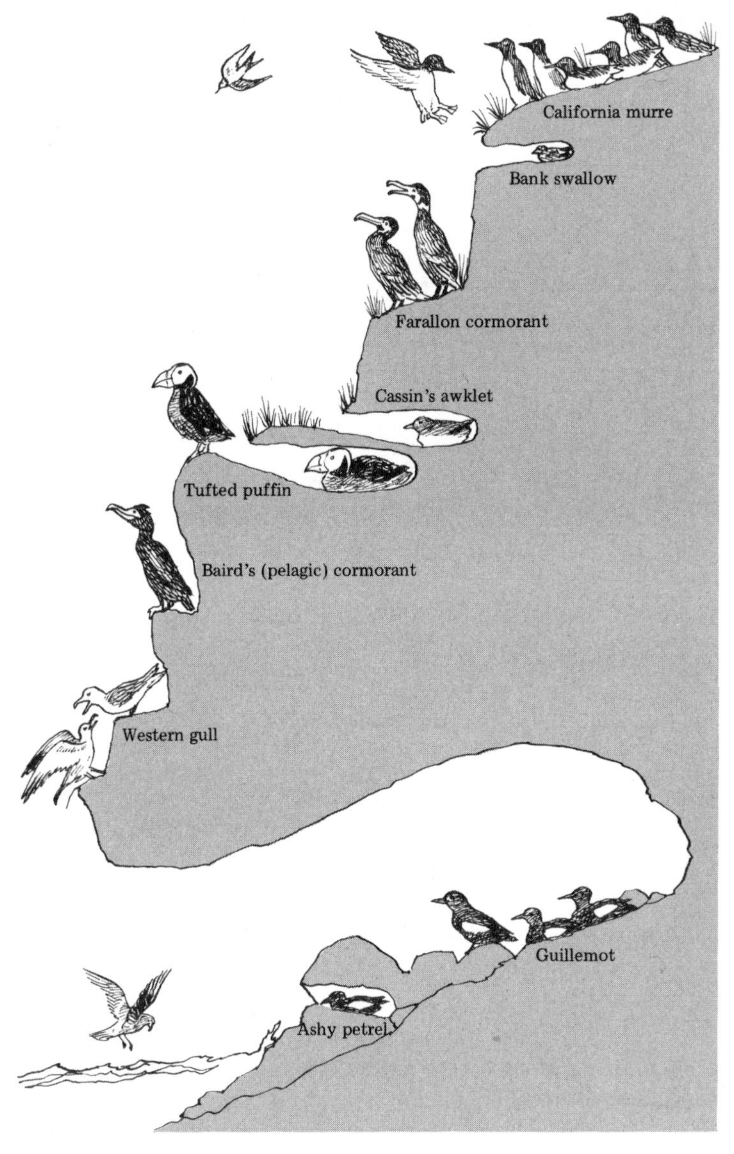

Fig. 1-3. Typical nest-site preferences of some California seabirds, showing niches and stratification of species in a rookery.

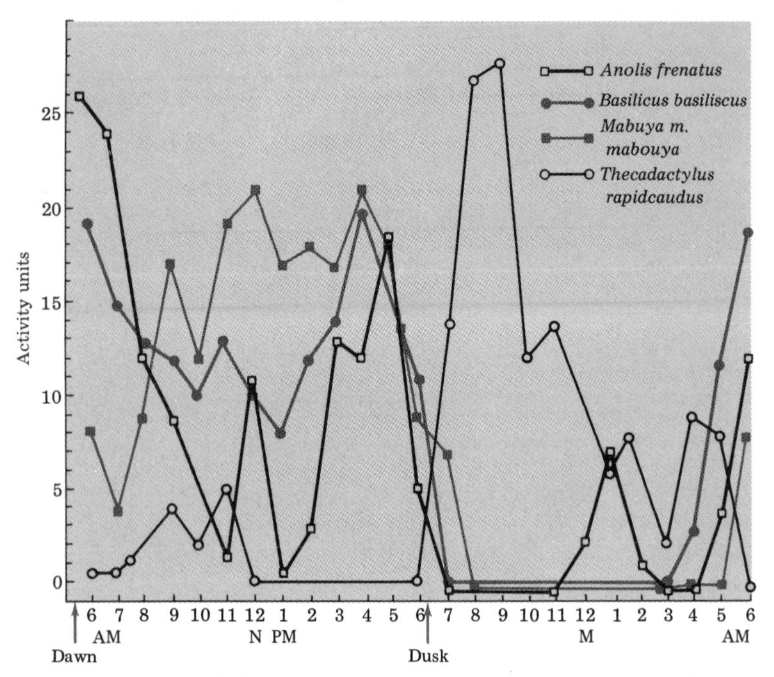

Fig. 1-4. The daily activity patterns of four Panamanian forest lizards. One species is a nocturnal gecko (*Thecadactylus rapidcaudus*); the other three species are diurnal. *Anolis* is an arboreal lizard similar to the Carolina "chameleon" sold in pet stores. *Basiliscus* is a large lizard that dwells along stream banks. *Mabuya* is a ground-dwelling, forest skink lizard. Each lizard species utilizes a different portion of the 24-hour day for its principal period of activity.

also transferred in orderly sequence within ecosystems. However, this transfer is *not* cyclical, because some energy is lost at every step and at the end of a chain of transfers. We shall consider these problems in more detail in Chapter 2. An understanding of these natural cycles will make the cycling and accumulation of such non-natural chemicals as pesticides immediately comprehensible. When we understand the relationships between these man-made chemical substances and natural cycles, we can better grasp their importance to man's present and future well-being.

Summary

Ecology is the scientific study of the interrelationships between living organisms and their environment, the physical world around them. It bears on every aspect of our lives, but is becoming more and more a vital topic to consider as our current population crisis grows in magnitude and our environment is further changed by man's actions. Ecology may be studied on the level of the individual, population, community, or ecosystem. Ecosystems, composed of biotic communities and their physical environment, exhibit evolutionary, spatial, temporal, and metabolic orderliness. The concepts of differential reproduction (natural selection), coevolution, ecological niche, habitat, adaptive strategy, stratification, and various types of periodism in activity are all necessary for an understanding of the structure and functioning of a biotic community.

2 SOME BASIC PRINCIPLES OF ECOLOGY: ELEMENTS AND ENERGY

In this chapter we consider in detail how chemical elements and energy flow through the ecosystem. Understanding these natural cycles thoroughly will help us understand such problems as how pesticides circulate in an ecosystem, the effects of mercury pollution, or the measures needed to capture more energy for food to feed man's expanding population on earth.

Biogeochemical Cycles in Ecosystems

Of the ninety-odd elements known to occur in nature, about forty are known to be required by living organisms. They are essential to maintain life. These chemical elements tend to be used over and over again, or to be *cycled*. They circulate in the **biosphere** (a term used to refer to all the living organisms on earth) in characteristic, more or less circular paths from environment to organisms and back to the environment again. Thus biogeochemical cycles involve pathways of elements moving repeatedly between inorganic forms and organic molecules. A summary of the biogeochemical cycles on our planet earth is given in Figure 2-1. It can be readily seen in this figure that there are several possible locations in the environment where elements may end up in inorganic form,

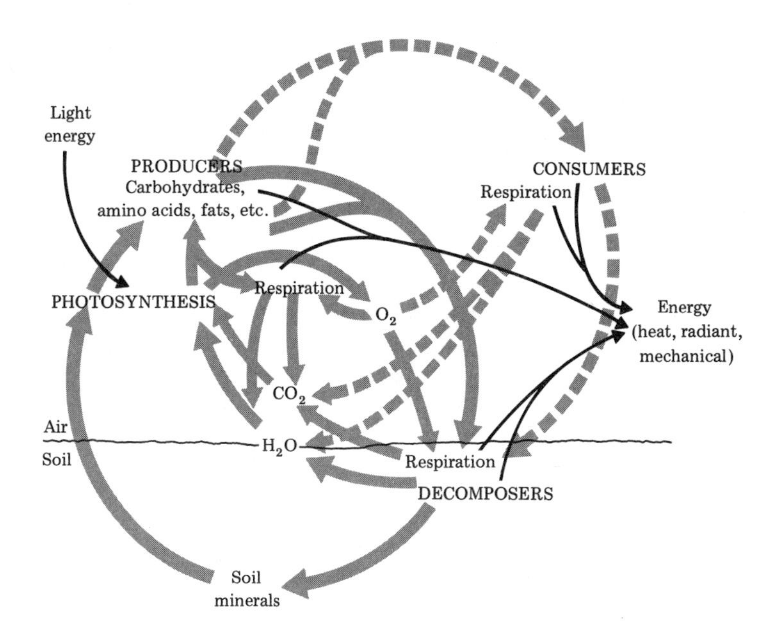

Fig. 2-1. Biogeochemical cycles on earth. Cycles of carbon, oxygen, hydrogen, soil minerals, and energy through the living organisms of the earth's biotic communities are indicated (nonbiological transformations are not shown). Solid lines show essential transformations between producers and decomposers. Broken lines indicate that consumers are not theoretically essential to a cycling of elements. Dotted lines show energy transfer from input as light energy to expenditure. Forked lines indicate that not all energy is lost through respiratory processes, but that some may enter the external environment via other metabolic and physical functions of the organism. Producers on earth are the photosynthesizers—the chlorophyll-containing plants. Consumers include the animals and many parasitic and saprophytic plants. Decomposers are primarily microorganisms, although it is sometimes technically difficult to draw a line between them and certain consumers. Water is indicated at the soil-air boundary because it usually enters metabolic reactions by first being taken up through plant-root systems.

and indeed the rate and "perfection" * of the cycle depend on which portion of the environment acts as a **reservoir** for the element. The cycles in which the element is returned to the environment as rapidly as it is removed by living organisms are said to be more perfect cycles than those in which part of the material is locked up in inaccessible chemical forms or geological formations for extended times.

Gaseous Cycles

These are more nearly "perfect" than other biogeochemical cycles in that the elements circulated do not become inaccessible to organisms over long periods. The gaseous cycles are the **carbon, nitrogen, oxygen,** and **hydrogen cycles.** These elements are moved about in tremendous amounts, with the earth's atmosphere serving as the main inorganic storage reservoir. While the four elements having gaseous cycles are only 10 percent of the forty essential elements, they constitute about 97.2 percent of the bulk of protoplasm (living matter) and cycle the most easily. We will examine two of these cycles briefly.

The **carbon cycle** (Fig. 2-2) is a good example of a "perfect" cycle. Carbon dioxide in the atmosphere (and dissolved in water) forms the principal inorganic reservoir of carbon. The photosynthetic reactions of green plants incorporate carbon dioxide from the air with water from the soil and through a series of chemical rearrangements produce organic compounds called **carbohydrates,** which include carbon atoms as a basic molecular "backbone." If the green plant should die, these organic compounds are broken down by the actions of decomposer organisms such as bacteria. As a waste product resulting from the respiration of decomposer organisms, carbon dioxide is immediately released back to the atmospheric reservoir. Should the green plant be eaten by an herbivorous or omnivorous animal, the carbohydrates and other carbon-containing molecules from the plant will be bro-

* Perfection in an ecological sense refers to the completeness of the cycling process, i.e., an element moving uniformly through all stages of the biogeochemical cycle without being held up for long periods at any point.

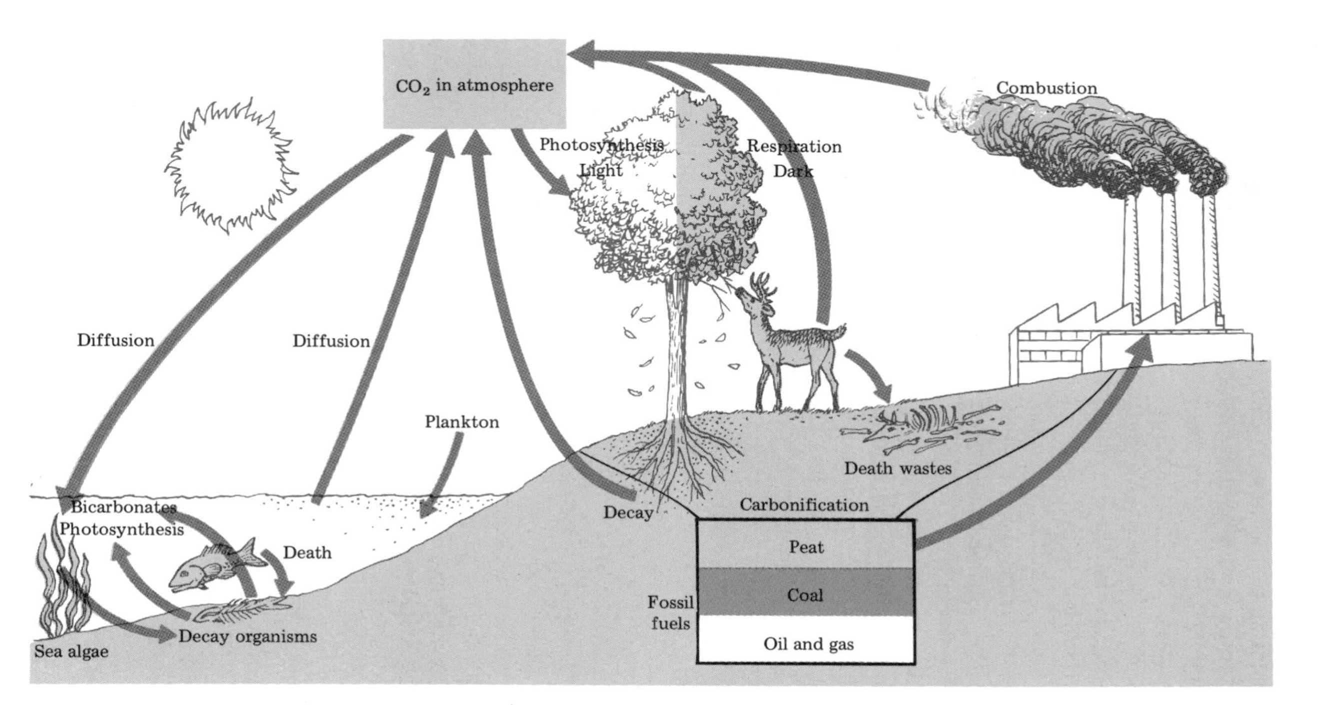

Fig. 2-2. The carbon cycle in an ecosystem. (*For credit see p. viii.*)

ken down in the animal's digestive system and body cells, and will eventually be synthesized into still different carbon compounds. Carnivores feeding on the flesh of these animals will break down and assimilate the compounds and re-use the carbon atoms in new organic molecules. In the end, though, respiration in these animal cells results in carbon being breathed out to the atmospheric reservoir in the form of carbon dioxide. Alternately, decomposer organisms break down the cellular organic molecules upon the death of the animal, releasing carbon dioxide to the atmosphere. Under certain conditions in past eras of geologic time, the carbon present in plants and animals was preserved in fossil fuels such as peat, coal, oil, and gas, and in rock formations such as limestone. Burning of the temporary fuel reservoirs and weathering of limestone return carbon to the atmosphere.

We can see by referring again to Figure 2-2 how important each part of the carbon cycle is. Should the green plants be destroyed, carbon would be unable to move from the inorganic atmospheric reservoir of carbon dioxide to organic compounds in protoplasm. If the decomposers in a biotic community were eliminated, organic matter from dead plants and animals would accumulate rapidly and the vital carbon atoms would be locked up, destroying the cycle.

The **nitrogen cycle** is another very complex, but more or less perfect, gaseous cycle (Fig. 2-3). About 78 percent of the atmosphere is nitrogen gas (N_2), the largest gaseous reservoir of any element. In its gas form, nitrogen is useless to most organisms. However, certain **nitrogen-fixing bacteria and algae** found in soils and wet habitats can convert inorganic N_2 into forms, especially **nitrates,** that are immediately usable by plants. Many of these nitrogen-fixing bacteria live in intimate association with **legume** plants (members of the pea family), in little **nodules** on the roots of clover, alfalfa, or other legumes (Fig. 2-4). The bacteria fix* nitrogen from the air by combining it with oxygen in the form of nitrates (NO_3), which the host plant or adjacent plants can absorb through their root tissue for use in protein synthesis. It has been estimated that for the biosphere as a whole one to six pounds of

* To fix in a chemical sense means to form a solid from the gaseous form.

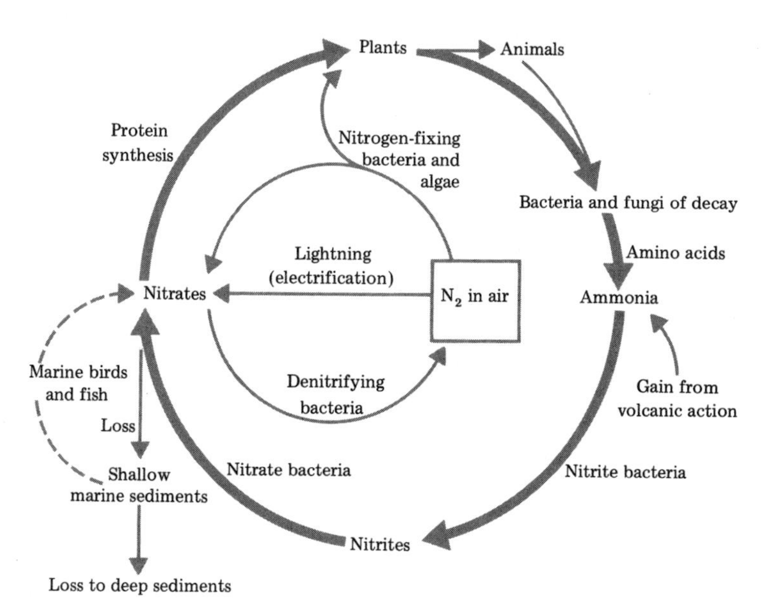

Fig. 2-3. The nitrogen cycle.

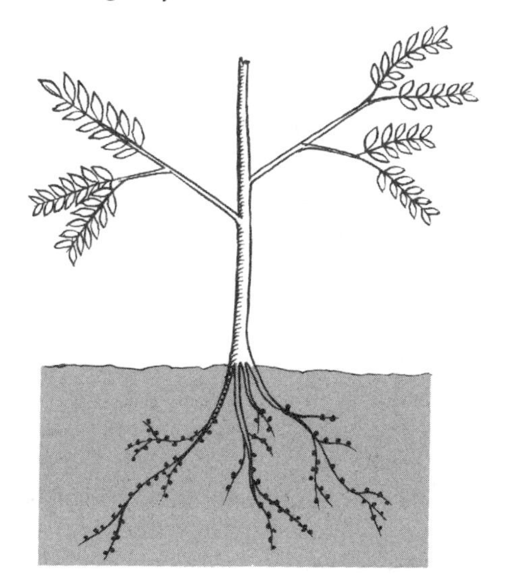

Fig. 2-4. Nodules containing nitrogen-fixing bacteria on roots of a legume plant.

nitrogen per acre per year are fixed from the air. In very fertile areas with a heavy concentration of legumes and nitrogen-fixing bacteria, up to 200 pounds of nitrogen per acre may be fixed! In water and moist soil, such as in oriental rice paddies, blue-green algae often perform the essential function of nitrogen-fixing. Increased yields result from adding extra algal cultures to these rice fields.

In the tissues of the plant whose roots absorb nitrates, the nitrogen atoms are used principally to synthesize amino acids, which are built into proteins. Several more organic rearrangements follow if the plant's protein molecules are passed through the digestive tracts and body cells of herbivorous animals and then (when herbivores are eaten) into the bodies of carnivores. When the plant or animal dies, decomposing bacteria and fungi cause the body to decay so that the nitrogen-containing amino acids are broken down, releasing ammonia gas (NH_3). **Nitrite bacteria** can convert this poisonous ammonia into simple nitrite (NO_2) molecules, and still other bacteria (**nitrate bacteria**) in the soil can add a third oxygen atom to nitrites to produce nitrates. At this point, we have gone full cycle, because plants in the area now have a usable form of nitrogen again. Note (Fig. 2-3) that nitrogen may be removed from the nitrates in the soil by **denitrifying bacteria** and returned to the atmospheric reservoir, from which it can be released again by either nitrogen-fixing bacteria or **electrification** by lightning. In the latter case, the energy of a lightning bolt passing through the atmosphere binds nitrogen and oxygen together into nitrates, which precipitate onto the soil from the air during electrical storms.

Thus the air is the greatest reservoir and safety valve of the system. Nitrogen is continually entering the air by the action of denitrifying bacteria. Nitrogen is continually returning to the cycle through the action of nitrogen-fixing bacteria or blue-green algae and through the action of lightning. Loss from the cycle occurs at the nitrate stage, when erosion and leaching of soils by rain wash nitrates into rivers and eventually into the ocean basins. Here they may be returned rather promptly to terrestrial ecosystems if marine birds and fishes that return to land or are brought there by man have fed on organisms utilizing nitrates from the shallow sediments near the shore. If nitrates in the shallow marine sediments are not absorbed by eelgrass, phytoplankton (microscopic free-floating plants), or other marine plants, they are gradually lost to deeper marine sediments where they re-enter the nitrogen cycle only upon the later geologic uplift of these sedimentary beds.

Sedimentary Cycles

These are said to be "imperfect" cycles because the constituent elements end up in sedimentary rock, a reservoir from which recycling is very slow. We have seen that carbon, nitrogen, oxygen, and hydrogen, constituting about 97.2 percent of the bulk of protoplasm, cycle quite easily because the principal reservoir in such elemental cycles is the gaseous form in the atmosphere. The rest of the approximately 36 elements comprising 2.8 percent of plant and animal tissues tend to go literally downhill in terrestrial ecosystems. Once they have been removed through erosion and other means of downhill transport, they have no immediate way of returning and hence their cycles extend across long periods of geologic time.

The **phosphorus cycle** is an excellent example of a sedimentary cycle (Fig. 2-5). Phosphorus is a necessary element in the hereditary material DNA, in other vital cellular molecules, and in the structure of bone in vertebrate animals. The principal reservoir for the cycle is phosphate rock formed in past geologic ages, although excrement deposits (guano) by fish-eating sea birds and fossil bone deposits contribute substantial phosphate in certain areas of the world. Erosion by rainfall and the runoff of streams dissolves phosphate out of these reservoir sources, forming a phosphorus "pool" in the soil. This makes phosphorus available to plants, which absorb it through their roots for use in cellular syntheses. Animals obtain phosphorus from plants; upon death or through normal excretion of waste products from the body, they return phosphorus to the dissolved phosphorus pool.

However, in the dissolved state much phosphorus is lost by downhill transport into shallow marine sediments. Some of this phosphorus is returned to land by sea birds who deposit

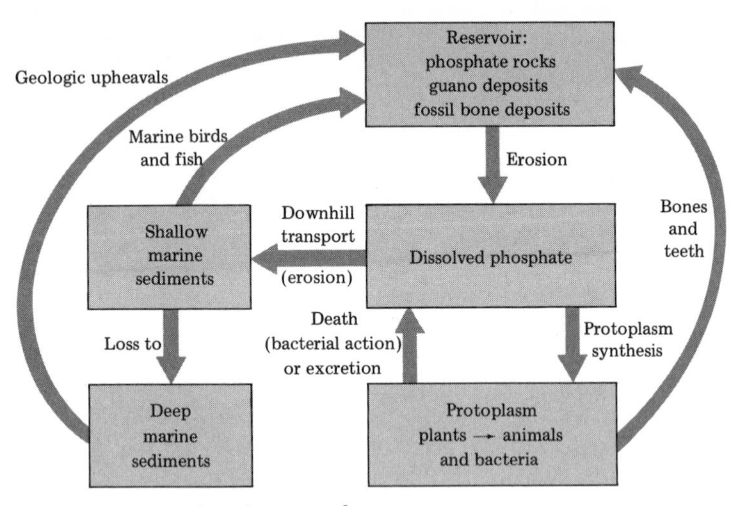

Fig. 2-5. The phosphorus cycle.

excrement on the shores and by migratory fish such as salmon and steelhead who carry phosphorus in their bones far inland on their return to fresh-water spawning grounds. These birds and large predatory fish have fed on smaller fish and other marine organisms, which in turn have fed on microscopic plankton. These plankton have incorporated phosphorus into their bodies from the shallow coastal sediments and estuarine waters where their early development normally takes place. But the majority of phosphate compounds is carried and lost to deep marine sediments by underwater currents and geologic subsidence and brought back to terrestrial reservoirs only through major geologic upheavals millions of years apart. The lack of sunlight for photosynthesis in the ocean depths, the low temperatures, and the high pressure there all prevent the extensive growth of plankton. Thus phosphorus on the deep ocean floor is simply not recycled before being covered by additional sediment and lost to the living world.

It appears to many experts that the present means of returning phosphorus to the cycle may soon be inadequate. There is no extensive upwelling of sediments currently in much of the world, and the transport function of formerly huge colonies of sea birds is not as extensive today as in past ages because of the depletion of many of these populations. Man does mine extensive amounts of phosphates from the earth, especially from central Florida, but the excessive use of these phosphorus materials as artificial fertilizers causes rapid loss through erosion and hence reduces the cycling effect still further.

How Man May Change Biogeochemical Cycles

One might think that the activities of man could have little effect upon such massive world-wide cycling of elements as goes on in the biogeochemical cycles just discussed. Up to recent times, this appeared true. Man did little to change the cycles except on a local scale; extensive deforestation and clearing might lower the rate of carbon or nitrogen cycling, or legume crops might be planted in rotation with more commercially valuable crops such as corn or cotton to restore nitrogen levels in the soil. However, we are now coming to realize that long-term effects on this almost inconceivably massive cycling may be occurring in ways we did not expect. A prime example of man's alteration of a major biogeochemical cycle is found in the carbon cycle.

Normally, the amount of CO_2 in the atmosphere is in equilibrium with dissolved CO_2 in fresh and salt water. Movement of the gas by diffusion from the air reservoir into the aquatic reservoirs equals the rate of diffusion of CO_2 back into the air from the water. From the CO_2 in the air or in water carbohydrate compounds are made by the interaction of sunlight and chlorophyll molecules in photosynthesis. Again, the amount of carbon tied up in protoplasm in this manner is in dynamic equilibrium with the CO_2 in the gaseous reservoir state. But over the past several hundred years man has been burning fossil fuels such as coal and oil at higher and higher rates. We are currently adding some six to nine *billion tons* of carbon per year to the atmosphere through the combustion of fossil fuels. Not too surprisingly, this is changing the composition of the earth's atmosphere. Let us look at the following data collected on the composition of the atmosphere since 1860:

Year	*Carbon Dioxide Content of Air (parts per million)*
1860	283 ppm of CO_2
1900	290 ppm
1960	330 ppm

In 100 years, the carbon dioxide concentration has increased an average of 47 molecules out of every million total molecules in the air. Currently we are burning between six and nine billion tons of fossil fuels per year, which would be sufficient to increase the amount of carbon dioxide in the air by 2.3 ppm per year. The measured rise of about 0.7 ppm, however, indicates that about two-thirds of the carbon dioxide released from fossil fuels is being dissolved in the sea or is increasing the total **biomass** (living weight of organisms) of land vegetation.

Even with this considerable increase, you might say CO_2 still represents only 0.03 percent of the atmosphere. Besides, more CO_2 would surely increase the rate of photosynthesis in green plants, which in turn would rapidly produce more useful oxygen. However, the amount of CO_2 in the air is critically important in maintaining the earth's temperature. Carbon dioxide, along with water vapor in the atmosphere, is translucent to the sun's electromagnetic radiation in the wavelength spectrum of visible light, but not to the longer infrared wavelengths. When solar energy (light) warms the earth's surface each day, infrared radiation (heat) is emitted back from the terrestrial sphere (Fig. 2-6). But CO_2 and water droplets do not allow these infrared wavelengths to escape, thus producing the so-called "greenhouse effect" on the earth's temperature. (In a greenhouse, the glass roof allows visible light to enter but traps the infrared radiation emitted from the interior surfaces and plants, causing a rise in temperature.) In the last hundred years there has been a 1.8° F rise in the mean temperature of the earth. This correlates precisely with the increasing amount of carbon dioxide in the air. This temperature trend, if continued, could potentially lead to the melting of the polar ice caps, which would raise sea levels drastically, to some 330 feet above present levels. Such states as Florida and most of the coastal areas around the world would be completely inundated. This predicted environmental effect gives us some idea of the importance of switching the energy sources of such major fossil-fuel consumers as power generating plants to atomic or fusion energy,

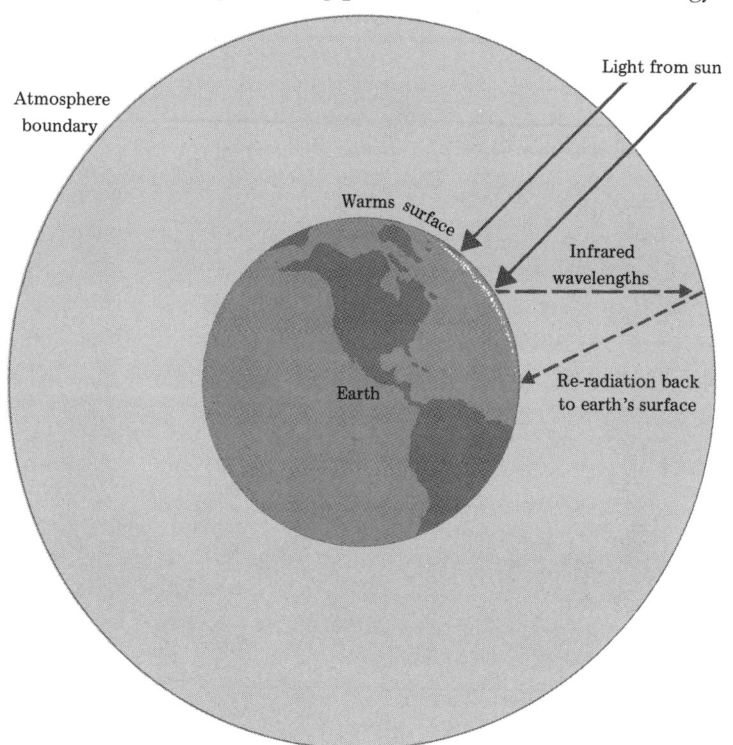

Fig. 2-6. The "greenhouse effect." Carbon dioxide and water vapor in the earth's atmosphere are translucent to the sun's electromagnetic radiation in visible wavelengths of light. These wavelengths heat the earth's surface and infrared wavelengths ("heat") are radiated back into the atmosphere. Carbon dioxide and water vapor, however, acting like glass in a greenhouse roof, do not allow these to pass. Hence the temperature of the atmosphere rises from the trapped infrared radiation.

provided that any environmental problems associated with these processes can be properly controlled.

Energy Flow in Ecosystems

Laws of Thermodynamics

In ecosystems energy is transferred in an orderly sequence, but energy flow is unidirectional, not cyclical. Only materials cycle. Some useful energy is lost as heat at every step in a chain of events, and two descriptive physical laws apply to this situation. These are the First and Second Laws of Thermodynamics.

The **First Law of Thermodynamics** deals with the conservation of matter and energy, and states that *energy cannot be created or destroyed but only changed from one form to another.* An example is given in Figure 2-7: visible light energy can be changed by the process of photosynthesis in green plants into chemical energy in the form of chemical bonds in the glucose sugar molecule. When the plant later uses this glucose molecule for food by breaking it up in the process of cellular respiration, the chemical energy present in the molecular bonds is released as heat, a third form of energy. Note that one form of light energy, infrared radiation, can result in heat energy too. The energy in the diagram is never destroyed, only changed from one form to another.

The **Second Law of Thermodynamics** states that some

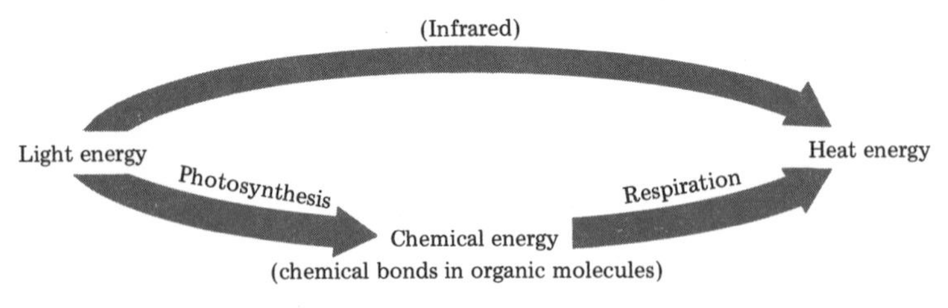

Fig. 2-7. Some relationships among three forms of energy.

useful energy (energy available for doing work) is converted into heat energy at every transformation of energy. This heat energy escapes into the surrounding environment. When you slide a box along the floor, some of the work you are putting into pushing the box is being used up as heat energy due to friction. In the same way, when you use energy stored in muscle cells to contract your arm muscles, some of the useful energy is lost as body heat from the arm's surface skin. In other words, energy transformations in the biological as well as physical worlds are less than one hundred percent efficient. Since heat energy cannot be used easily to do work, more energy must be supplied to a biological system from outside the organism to counterbalance the inevitable loss of energy as heat. In fact, the universe as a whole is running down; it is tending towards a state of maximum **entropy,** a condition of random disorganization where all energy will be transformed into heat energy at a uniform temperature. In order to continue to function, organisms must continue to receive new supplies of energy in the ecosystem. Ultimately this energy comes from outside our earth in the form of light energy from the sun. The passage of energy through an ecosystem may be diagrammed as:

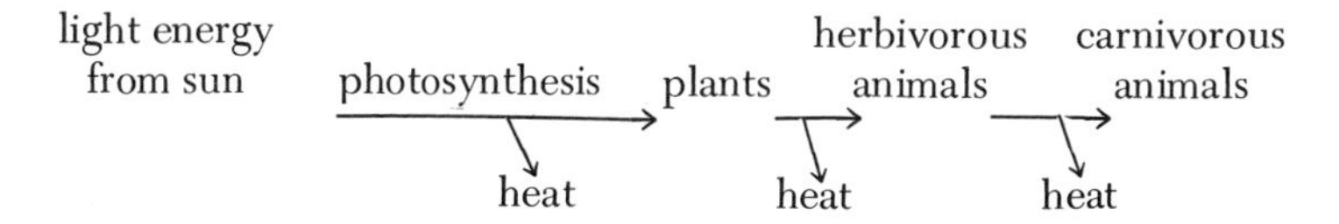

Energy has to be added constantly to the ecosystem via solar radiation. Plants convert this light energy into chemical bond energy during the process of photosynthesis, but much of the original energy is lost as heat (shortly in this chapter, we shall see just how inefficient this conversion into chemical energy really is). When plants are eaten by herbivorous animals these chemical bonds are broken and then reformed into new molecular bonds within the animals' cells. More energy is lost as heat at this stage too. When carnivores eat the herbivores, still another sequence of breaking and reforming of chemical bonds occurs, and further heat energy escapes unused by the carnivore.

Given these principles of energy transfer, how can we more precisely describe energy flow between the component species of an ecosystem? The simplest form of this flow is through the **food chain.**

The Food Chain Concept

A food chain simply transfers food energy from a given source through a series of species, each of which eats the one before itself in the chain. This repeated series of eating and being eaten is always initiated with green plants, which receive their energy from the sun. A very simple food chain is:

grass ⟶ cow ⟶ man

heat energy lost heat

At each transfer a large proportion of the potential energy present in the chemical bonds of the food species is lost as heat. This limits the number of steps in a food chain, usually to four or five. In the aquatic ecosystem at Silver Springs, Florida, for example, one of the longer natural food chains is:

algae (microscopic green plants) ⟶ midge (fly) larvae ⟶ bream (fish) ⟶ bass

We shall see shortly and in more detail how the amount of energy at each step limits the number of links in this chain.

A final attribute of food chains is that the shorter the food chain (or the nearer the organism is to the beginning of the chain), the greater the available energy that can be converted into biomass (living weight) and utilized in cellular respiration. In a two-step food chain of

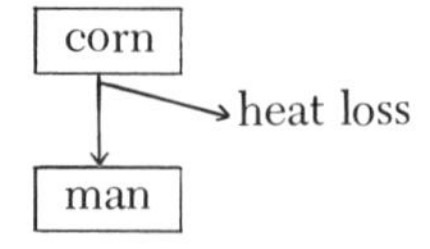

more of the potential energy initially available per pound of corn reaches man than if the corn is first fed to pigs and the swine are then eaten by man, as in this three-step food chain:

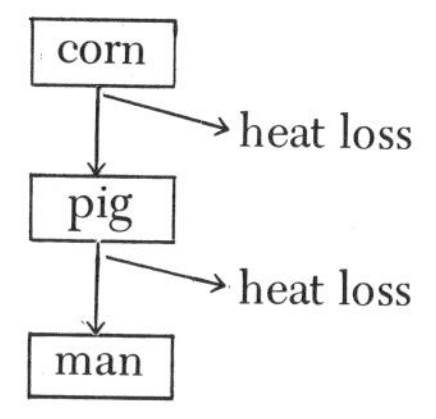

Normally we visualize food chains of the **predatory** type, where smaller organisms are eaten by larger organisms, which in turn are eaten by still larger species. Some food chains involve larger organisms being eaten by smaller species; these are the **parasitic** food chains. An example would be a deer fed upon by internal roundworms or external ticks, or a man with malarial parasites in his blood. A third special term, **saprophytic,** is given to a food chain where dead organic matter is fed upon by fungi such as mushrooms and finally bacteria. (Saprophytes are any plants that feed on dead and decaying matter.)

If we had only a series of isolated food chains to deal with in nature, the transfer of energy in an ecosystem would be very simple indeed. However, we find in any biotic community that these food chains are interlocked in a complex pattern of feeding relations that we call a **food web.**

The Food Web Concept

A food web is simply the total set of feeding relationships in a biotic community. With many interlocking food chains the community will remain stable even if one or more of these relations is altered. At Silver Springs, for instance (Fig. 2-8), we see that if one of the food species (e.g., the bream) for the bass should become rare or absent (through heavy predation by another organism, seasonal migration out of the area, and so forth), the bass are not forced to die or move out. They can feed more on the other species (mullet, caddisfly larvae, water beetles) that are in the same nutri-

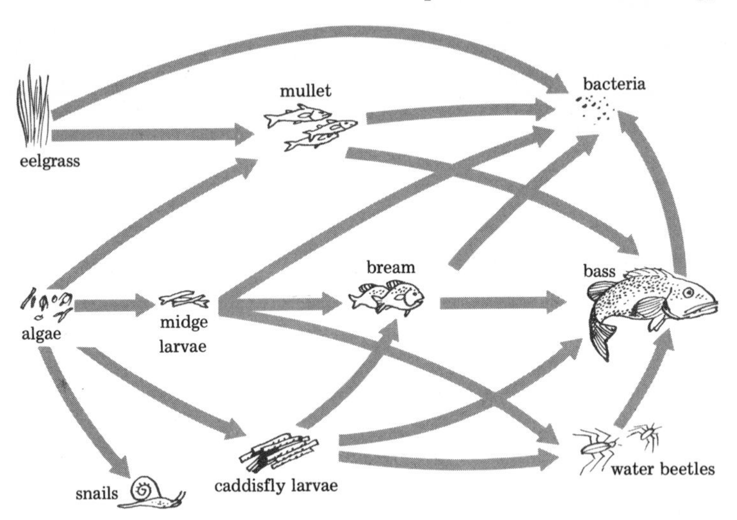

Fig. 2-8. Part of an aquatic food web at Silver Springs, Florida.

tional relationship to them as the bream. Obviously, then, a food web introduces a strong element of *stability* into an ecosystem. The more components involved, the more stable the system will probably be.

The Trophic Level

All organisms that share the same general source of nutrition are said to be at the same **trophic level.** This feeding level concept implies that these organisms obtain food through the same number of steps from plants in a food chain. The first trophic level (Fig. 2-9A) in ecosystems comprises the **producers,** organisms that photosynthesize (convert light energy from the sun into chemical-bond energy)—in other words, green plants. The **herbivores,** animals that eat plants, are the second trophic level (Fig. 2-9B). (Because herbivores are the first organism level to feed on other organisms, their trophic level is sometimes called the primary consumer trophic level.) The **carnivores,** animals that feed on herbivores, are the third trophic level (Fig. 2-9C) or secondary consumer trophic level. The **top carnivores** are animals that feed on other carnivores; they represent the fourth and sometimes even the fifth trophic levels (tertiary or quaternary consumers). The **decomposers** (Fig. 2-9D), organisms that feed upon dead organic matter, consume plants and animals from all trophic levels. They perform the instrumental role of breaking down dead plant and animal bodies, reducing them finally to inorganic soil material. Their action breaking down the fallen trunk of a pine tree (Fig. 2-9E) illustrates the sequential scheme that decomposition often follows. Large tunneling insect species such as beetles begin the gross breakdown of the wood structure. These are followed by carpenter ants, and eventually termites find the decomposing wood suitable for their tunnels. Fungi grow in the ever moister wood. Eventually bacteria perform the ultimate decay functions and reduce all the material into organic molecules, which can dissolve into the soil.

You are already probably thinking from the above discussion that a species *can* occupy more than one trophic level. Man does it daily in most parts of the world, eating meats and grains simultaneously. A bear (*Ursus*) feeds on plant material such as berries and roots (producers) and then may catch animal prey such as mice or ground squirrels (herbivores). This bear is in the second and third trophic levels simultaneously.

Returning to our original overall problem of energy transfer, we may now ask: what is the pattern of energy flow in an ecosystem in terms of these trophic levels? The general observation we can make in every ecosystem is that out of the tremendous quantities of solar energy falling on an area each year, only a tiny fraction of the light energy is actually utilized by the plants and animals living there. On the average in ecosystems around the world, more than 98 percent of the available light energy from the sun is lost and less than 2 percent is trapped by plants and stored as chemical-bond energy. At each successive transfer between trophic levels, more energy is lost via respiration and heat. This happens in aquatic ecosystems as well as terrestrial systems (Fig. 2-10). The inexorable principles which we saw expressed in the laws of thermodynamics insure that plants and animals simply are not able to utilize all the energy available to them. Figure 2-11 summarizes this general pattern in a crudely quantitative way. This can be done much more accurately by

A

Fig. 2-9. A. The first trophic level: producer organisms in a tropical rain forest in Costa Rica. B. The second trophic level: herbivores on the African plains. C. The third trophic level: predatory spider (carnivore) feeding on a tropical heliconiine butterfly. D. The fate of all trophic levels is shown in this photograph of a decomposer at work—a tropical fungus on a dead fly. E. Reduction of a fallen pine log to inorganic soil by a succession of various decomposer organisms.

B

C

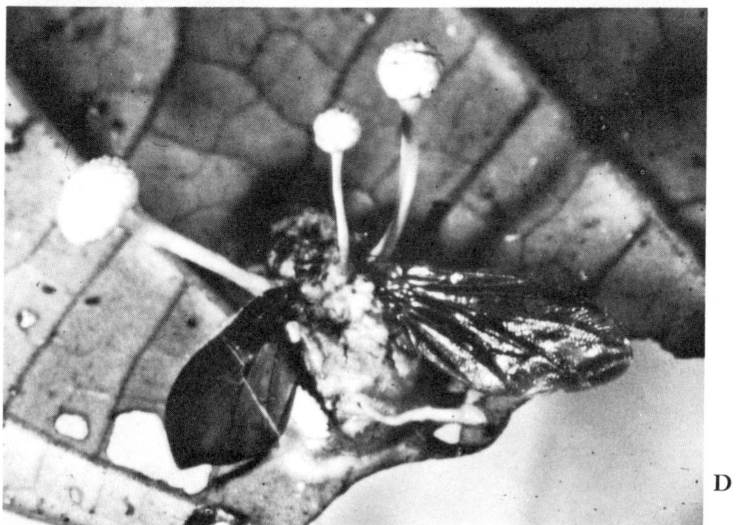

D

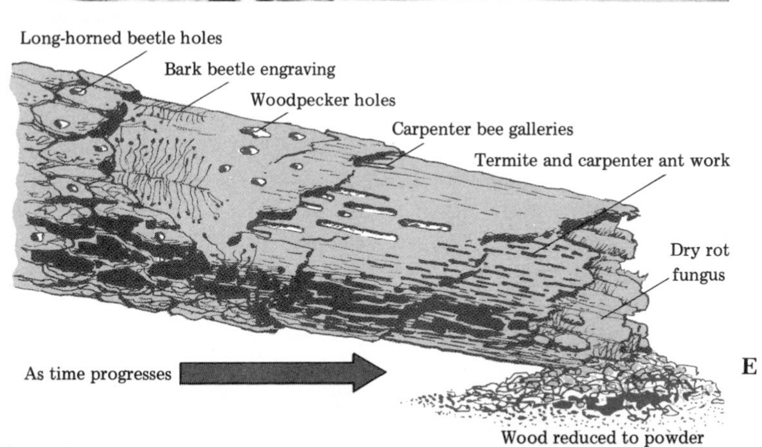
Long-horned beetle holes
Bark beetle engraving
Woodpecker holes
Carpenter bee galleries
Termite and carpenter ant work
Dry rot fungus
As time progresses
Wood reduced to powder

E

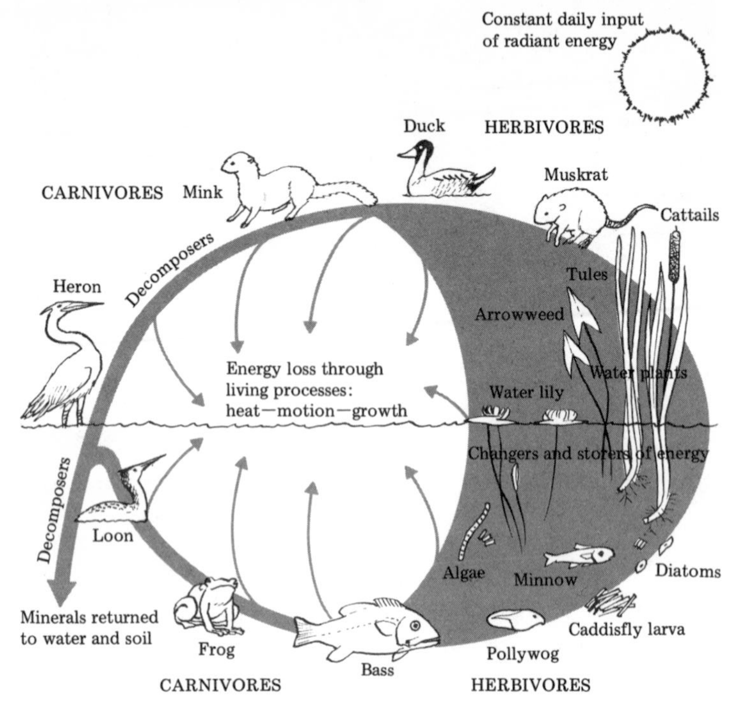

Fig. 2-10. Typical energy flow patterns through a freshwater marsh.

Fig. 2-11. Diagrammatic representation of energy loss in a typical ecosystem on earth. Sunlight enters from outside the system and most light energy passes through, with less than 2 percent trapped by photosynthesis.

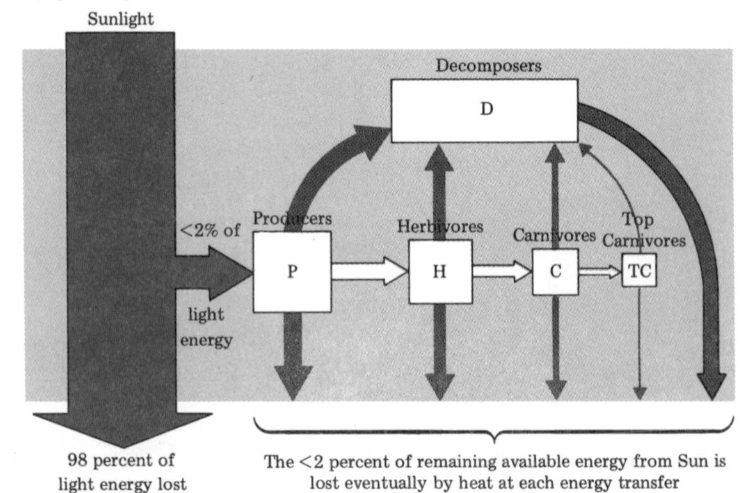

calculating the ecological efficiencies of these trophic levels, which we shall now consider.

Quantifying Energy Flow in Ecosystems

We would like to be able to determine for any ecosystem, whether an agricultural field crop or a mountain forest, how much of its potential food supply each trophic level is using. In other words, how efficient is a particular level in exploiting the total amount of food available to it? This is a very practical kind of ecological question because if we know how efficient a level is in a particular environment (say, the corn plants in an Iowa field), we may be able to manipulate the environment in certain ways to increase that efficiency for the ultimate benefit of man.

Ecological efficiencies are calculated as the ratios between energy flow at different points along the food chain. An ecologist named Howard T. Odum did a classic study on the Silver Springs, Florida, ecosystem in which he measured how many units of energy were captured by each trophic level over a year's time. The results of his study are depicted in Figure 2-12. At this particular latitude and locality, Odum found that 1,700,000 kilocalories (units of energy *) were falling on each square meter of Silver Springs over a 365-day year. Of this amount, only 20,810 kcal of energy were captured by green-plant producers through photosynthesis in each square meter per year. Less than 2 percent of the available solar energy was trapped by chlorophyll molecules ($\frac{20,810}{1,700,000}$ kcal by our definition of the ecological efficiency of the producer level). Over the course of a year, herbivores consumed and utilized some 3,368 kcal out of the 20,810 kcal available in the plant material in each square meter: a 16 percent efficiency (84 percent was lost to the environment). And we can continue up through the pyramid diagram, determining the efficiency of energy utilization for each trophic level. Do note in particular that the decomposers are quite ef-

* One kilocalorie (kcal) represents the amount of heat required at a pressure of one atmosphere (sea level) to raise the temperature of one thousand grams of water one degree centigrade.

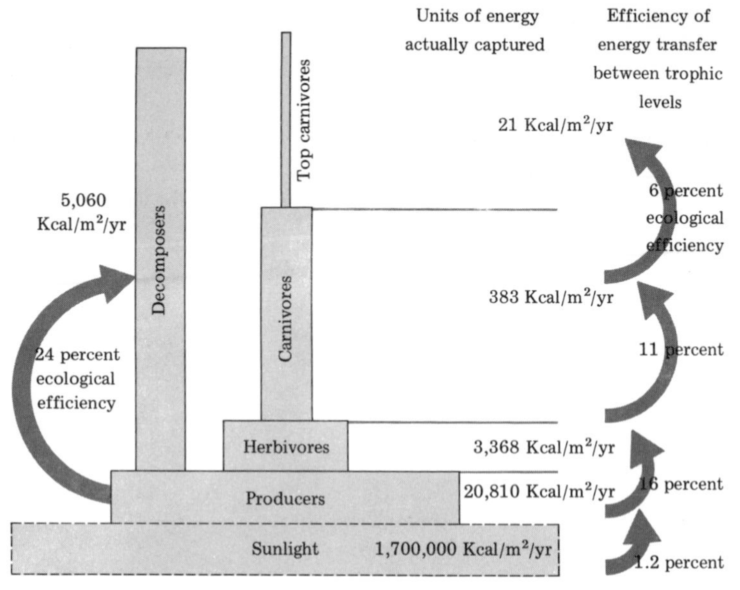

Fig. 2-12. The ecological efficiencies of the trophic levels at Silver Springs, Florida.

ficient. Feeding on all trophic levels, they utilize some 24 percent of the original potential energy available in the chemical bonds of the producer plants.

Table 2-1 summarizes the efficiency of energy transfer at various trophic levels in three aquatic ecosystems. Note the very *low primary efficiencies* in the plants, on the order of 1 percent or less. Apparently very low efficiencies are necessary for maximum output by plants. Photosynthetic plants have only a few hours of high sunlight intensities to operate in each day; so apparently rapid growth per unit time (with accompanying waste of energy) is more valuable for survival than maximum efficiency in the use of fuel (solar energy). Plants in suitable areas *could* grow more slowly and be more efficient in trapping and using energy, but they do not. Perhaps this is because of high competition with other plants (in the tropics) or short growing seasons (in the temperate zone and deserts).

Also note that there is generally *increasing efficiency* in higher consumer groups. Plants cannot move much to capture

TABLE 2-1. EFFICIENCY OF ENERGY TRANSFER AT VARIOUS TROPHIC LEVELS IN THREE AQUATIC ECOSYSTEMS (AFTER ODUM, 1959)

	Trophic level energy intake efficiency (percent of available energy that a trophic level acquires)		
Trophic Level	*Cedar Bog Lake, Minn.*	*Lake Mendota, Wisc.*	*Silver Springs, Fla.*
Photosynthetic plants (producers)	0.10	0.40	1.2
Herbivores (primary consumers)	13.3	8.7	16.0
Small carnivores (secondary consumers)	22.3	5.5	11.0
Large carnivores (tertiary consumers)	not present	13.0	6.0

sunlight; hence great amounts of light energy never hit the leaves or photosynthetic cells. Herbivores can move around to feed, and the searching behavior of predators greatly increases their chances of finding suitable prey.

Ecological Pyramids

We have seen that in a food chain there is energy loss at each transfer, and that this involves a regular sequence of trophic levels. This picture of trophic structure can be summarized graphically in the form of an **ecological pyramid** (Fig. 2-13). The first or producer level forms the base, and successive levels form the tiers which make up the apex. Ecological pyramids or food pyramids may be of three general types:

1. the *pyramid of numbers,* in which the number of individual organisms is depicted for each trophic level;

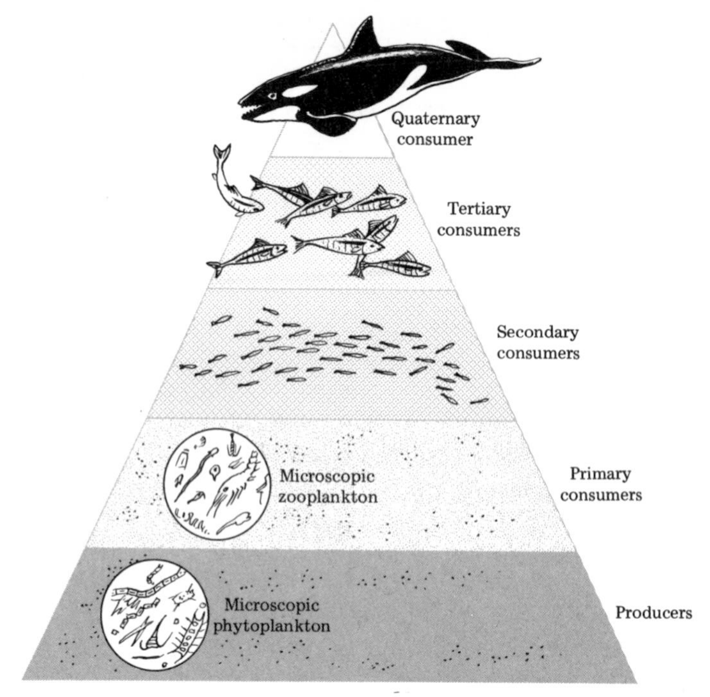

Fig. 2-13. An ecological food pyramid in the sea.

2. the *pyramid of biomass,* based on the total dry weight or caloric value or other measure of the total amount of living material in each trophic level;
3. the *pyramid of energy*, in which the rate of energy flow and/or productivity (rate of production) of successive trophic levels is shown.

Why should a pyramid be used to depict energy relationships in a community? First, of course, whether we are dealing with nature or geometry, many small units are required to equal one large unit. One larger animal needs to feed on many smaller ones of a lower trophic level in order to secure sufficient energy for its own metabolic requirements. Second, there is a loss of energy at each step in a food chain. Thus there is simply less energy available to the higher trophic levels and they have to eat more to obtain enough for survival.

Summary

Biogeochemical cycles maintain a constant movement between physical environment and biosphere of the approximately forty chemical elements known to be essential to life. Gaseous cycles, with the atmosphere forming the chief reservoir, are said to be perfect cycles because the elements involved (carbon, nitrogen, oxygen, or hydrogen) do not become inaccessible to organisms over long periods. Sedimentary cycles are imperfect cycles because the elements involved end up in sedimentary rock, a reservoir from which recycling is very slow. Man's recent activities are causing significant changes in natural cycles such as the carbon cycle.

Energy flow in ecosystems, unlike the flow of elements, is unidirectional and not cyclical. Energy (as heat) is constantly lost in the process of transfer of energy, as expressed in the First and Second Laws of Thermodynamics. A food chain is a simple representation of this flow of energy through the feeding relationships of a series of species. A food web, composed of many interlocking food chains, more accurately represents the pattern of energy relationships in a community. All organisms that share the same general source of nutrition are said to be at the same trophic level; these trophic levels in a community are the producers, herbivores, carnivores, top carnivores, and decomposers. Ecological efficiencies are a method of quantifying energy flow between trophic levels. They show that higher trophic levels tend to be more efficient than lower ones at utilizing available energy. Trophic structure may be graphically represented by an ecological pyramid, which can refer to numbers, biomass, or energy present at each trophic level in the community.

3
SOME BASIC PRINCIPLES OF ECOLOGY: INTERACTIONS OF ORGANISMS

Great fleas have little fleas upon their backs to bite 'em,
And little fleas have lesser fleas, and so *ad infinitum*.

Augustus de Morgan (1872)

Organisms, whether they live in a terrestrial or aquatic community, interact with each other constantly. As we saw in Chapter 2, many of their contacts result in energy flow through the ecosystem. On a long-term basis, these interactions result in ecological and evolutionary changes in one or more of the species involved. The three basic types of interspecific relations in a biotic community are **predation, symbiosis,** and **competition.** Understanding the ecological outcomes of these relationships will help us see how communities are structured; then, in the following several chapters, we will study the organization and dynamics of the populations of the individual species that make up a biotic community. In other words, we will move from the global scale of biogeochemical cycles in ecosystems to energy flow in a particular ecosystem to community relationships and finally to the organization of a population. Let us now consider the basic interspecific relations.

Predation

Predation may be defined as the behavior of capturing and feeding on another organism with the latter being consumed wholly or in part. This has a negative effect on the growth and survival of individual members of the population who are being eaten. The two components of a predation system are the predator and the prey species. The **predator** is a free-living organism that feeds on other living organisms, usually of another species. Intraspecific predation, more commonly called cannibalism, is almost always associated with conditions of overcrowding in a population. The **prey** is an organism that is eaten by a predator. Both carnivorous and herbivorous animals are predators (the latter eat plants as prey).

We are usually more concerned with animals in thinking about predation, but, surprisingly, there are also predatory plants. Among these are such unique species as the Venus' Fly-trap (*Dionaea muscipula*) and the aquatic bladderworts, which capture living prey in a leaf that closes like a steel trap. The Sundews (*Drosera* species), found in many boggy parts of the United States, have sticky secretions at the tips of many bright red hairs standing above the paddle-shaped terminus of a leaf. When unwary insects land to investigate the bright color and glistening droplets, they become firmly stuck and the sensitive hairs turn inward, enfolding the prey in the center of the leaf blade. The plant cells of the trap then secrete digestive enzymes that break up the bodies of the prey. The chemical molecules that formerly composed the prey are then absorbed into the cells lining the trap.

Pitcher plants use the pitfall method to capture their prey. An unwary insect descends into the deep vertical trap cavity and finds hundreds of downward-pointing hairs blocking its return. It eventually falls into the fluid in the bottom of the trap, dies, and its proteins are digested by both plant-secreted enzymes and bacterial action. The lassoing fungi in water or wet soil are able to trap nematode worms unwary enough to pass through a loop of fungus cells. Examples of some of these predatory plants are shown in Figure 3-1. Carnivorous plants, however, are relatively rare in the plant

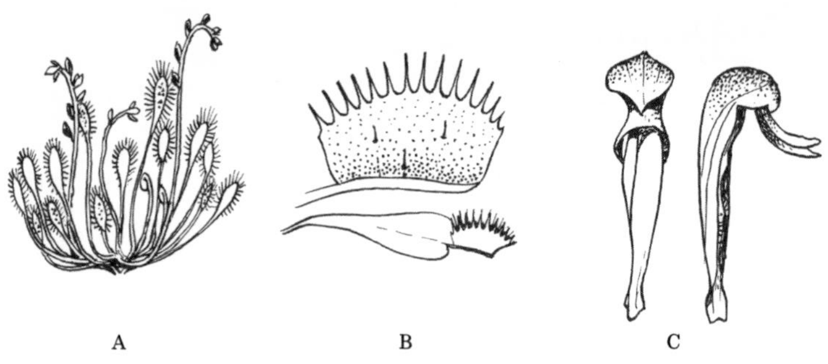

Fig. 3-1. Some carnivorous plants found in the United States. (A) A Sundew (*Drosera*), which preys on tiny insects by catching them with sticky globules on the ends of the hairs on its leaves. (B) The trap and the trap with leaf petiole of a Venus' Fly-trap (*Dionaea muscipula*), native to North Carolina. The three trigger hairs on the inner surface of each trap lobe will cause the trap to snap shut in about 0.3 seconds when touched by an insect prey. (C) A pitcher plant (*Darlingtonia californica*) from the West Coast. This particular species has a hood covering most of the open mouth of the "pitcher."

kingdom and we may mainly regard them as curious anomalies and evolutionary specialists rather than as a major kind or class of predator.

Animal predators are commonplace to us. Hawks (Fig. 3-2) and other predatory birds are still seen even in suburban areas of the United States. Man is as much a predator as a house cat or dog, though we usually do not see our food killed and processed when we purchase it. Plains Indian tribes were well known among frontiersmen in western North America as expert hunters who could bring down even a three-thousand-pound buffalo with bow and arrow (Fig. 3-3). Today our predation is done for us by the slaughterhouse and the farmer harvesting his vegetable or cereal crops.

We tend to picture predation as harmful to the prey species. Though on an individual basis we can say predation is destructive in character, over the long term it is almost always beneficial to the prey population as a whole. Later we shall consider in detail the ways in which predation helps regulate population sizes in both animals and plants. For now let us note that the effect of predation interaction on the

Fig. 3-2. A sharp-shinned hawk with a freshly captured sapsucker.

Fig. 3-3. Man was a direct predator during almost all of his tenancy on earth.

prey population will differ in degree depending upon how long the species have been associated with each other. A predator-prey relationship of recent origin usually has severe negative effects on the prey population; in other words, large numbers of the prey species are killed by the predator. New predator-prey relationships can occur when there is a major alteration in the ecosystem (often produced by man), or when a predator or prey species is first introduced into a region. Either event sets up a new frame of reference for the organisms concerned. The species in that ecosystem begin adjusting and determining a novel set of biological relationships and relationships to new physical environmental factors. Species that were not previously preyed upon by a particular predator may now become the chief source of food for it.

In a predator-prey relationship that has existed for a long time, **coevolution** (essentially simultaneous evolutionary adjustments in each of several species) minimizes the negative effects on the prey. In other words, if adjustments in the interaction were *not* made by natural selection over the course of time, one or both of the populations would probably become extinct. The prey would be eaten into extinction and the predator species would die of starvation unless alternate prey species became available. Instead, in continuing predator-prey relationships we generally have a dynamic balance between predator and prey. Adjustments in such factors as the birth rates and dispersal of each minimize the disturbance of the system and continue the status quo.

Several widely observed aspects of animal coloration have arisen through natural selection in response to predator-prey interactions. In addition to sharp claws, canine-like teeth, rapid reactions, and other prey-capturing adaptations, many vertebrate predators have color lines leading forward from the eye that apparently function as aiming sights. These circles and lines are found about the eyes of predatory birds, mammals, reptiles, amphibians, and fish (Fig. 3-4). Besides their role as lines of sight in tracking and capturing prey, dark eye marks also reduce glare in bright open habitats. Light circles around the eyes probably function as light-gathering devices (Fig. 3-5). Man has adopted these principles in the sphere of athletics by applying dark grease or charcoal smears just below the eyes of players to reduce the glare caused by sunlight or artificial lighting.

Another evolutionary result of predation pressure is **protective coloration,** where the prey comes to resemble some

Fig. 3-4. Eye lines serve as sighting and aiming devices for predators in catching prey. (A) Simple eye line of the partially insectivorous blue tit (*Parus caeruleus*). This is the most common type of eye line in vertebrates. The eye line in this species is wider than it is in other avian examples. (B) Combined eye circle and eye line of the yellow-throated vireo (*Vireo flavifrons*). (C) Raised yellow feathers above the eye line in a white-throated sparrow (*Zonotrichia albicollis*); these may cast light along the line of sight. (D) Red-necked grebe (*Podiceps griseigena*) showing eye line slanting downward. Such a line also occurs in some other fish-eating birds. (E) Long-billed curlew (*Numenius americanus*) showing direction of eye line forward of center of pupil to bill tip. (F) Teardrop mark of the pickerel (*Esox americanus*) associated with downward dashes at prey. (G) Rearward pointed eye line of the European woodcock (*Scolopax rusticola*), probably used to sight predators coming from behind. Associated with 360° vision plus front and back binocularity. (H) Head of the arboreal vine snake (*Oxybelis aeneus*) showing eye line and groove.

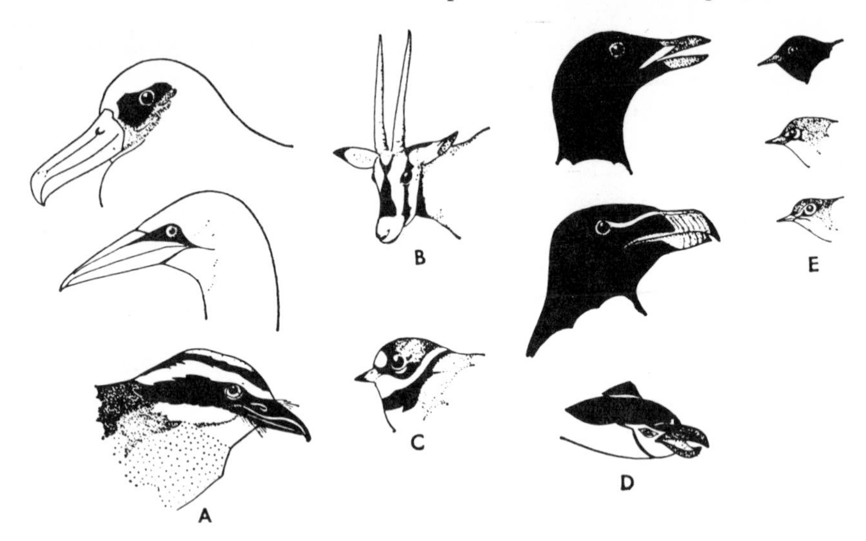

Fig. 3-5. Figures showing facial markings with adaptive functions other than predatory sighting alone. (A) Facial patches and stripes associated with reduction of glare in very bright habitats. (B) Here eye lines may hide eyes and further decrease visibility to potential predators by interrupting the beast's outline; also, the pattern may decrease glare on the line of sight toward clumps of grass in the animal's usually bright open habitat, and the pattern may be important as a social signal. (C) Here there is no sighting line; stripes are probably mainly for disruption of the head pattern (hiding the eye) and for aiding species recognition. (D) Alcids and penguins are very highly specialized for catching fish. The lines about the eyes and heads of virtually all birds of these two groups could not be for sighting on prey; they probably serve in intra- and interspecific recognition. (E) Male (top), female (middle), and first-year male (bottom) American redstart (*Setophaga ruticilia*), showing development of differences for recognition of sex, species, and age as well as a complex eye line and eye circle associated with feeding.

object in its environment, thereby deceiving potential predators. Protective coloration has profound ecological as well as evolutionary implications, for many more prey species can be packed into a community if each is protectively colored in some way, reducing the impact of predation upon the individual species. Several examples of types of protective coloration are shown in the accompanying figures.

In **cryptic coloration** (Fig. 3-6) the prey resembles or mimics some inanimate object or background. This type of protective coloration is commonly found among insects, such as walking sticks, "stick" caterpillars of geometrid moths, leaf-like katydids, or beetles that resemble bark or rocks.

In **Batesian mimicry** the prey resembles a toxic, venomous, or unpalatable species found in the same region. The tasty or at least harmless prey is the **mimic** and the species dangerous to potential predators is the **model** in a Batesian mimicry complex. Behavior is an important component of the mimicry, as otherwise the deception provided by similar external structure, coloration, and pattern would not convince the predator that it was seeing an unsuitable prey.

The monarch butterfly (*Danaus plexippus*) is a common species in most parts of the world and is distasteful to birds, lizards, and other predators. The adults are large slow-flying butterflies with bright and showy red-orange coloration bordered with black (Fig. 3-7). The adult is poisonous because of the diet of its caterpillar. As it feeds upon the leaves of the milkweed foodplants, the larva incorporates into its body poisonous compounds called cardiac glycosides. When the larva pupates and the body tissues reorganize inside the chrysalis (pupa) to form the adult, these compounds are not broken down but are built into the tissues of the adult monarch. When the adult emerges from the pupa, it is recognized as a distasteful insect by predators with previous experience of the species. In the United States, the monarch is the model in a Batesian mimicry complex with the viceroy butterfly, a good mimic. The viceroy belongs to an unrelated group of butterflies, but the adult has evolved to fly and look almost exactly like the monarch. Thus, although it is edible (the viceroy larva feeds on willow and similar non-poisonous plants), potential predators such as birds will shy away from it, mistaking the adult for a monarch adult.

Several kinds of **repelling coloration** exist, where the coloration of the prey warns predators of its genuinely noxious qualities. In **Mullerian mimicry,** for instance, all potential prey species in a mimicry complex are distasteful and share a common warning color pattern and behavior (Fig 3-8). This is a survival advantage to each of the prey species because

A

Fig. 3-6. Protective (cryptic) coloration shown by: (A) boa constrictor on rain forest floor; (B) a tropical *Hamadryas* butterfly on a tree trunk; (C) a walking-stick insect in Colorado; and (D) two poor-will nestlings (family Caprimulgidae) in bare "nest" on forest floor in Costa Rica.

B

C

D

Fig. 3-7. Batesian mimicry complex in the eastern United States: the Monarch (*Danaus plexippus*) is the distasteful model and the Viceroy (*Limenitis archippus*) is the edible mimic.

Fig. 3-8. Mullerian mimicry complex in the Neotropics (Costa Rica): all four specimens are different species of distasteful *Heliconius* butterflies (*H. hewitsoni, pachinus* [left]; *sarae* and *doris* [right]). On structural grounds it is believed they are not at all closely related species, but have converged solely in color pattern.

the predators in the area will learn to associate distastefulness with that pattern without trying all the species. A predator has only to try one species of yellow- and black-banded wasps to know that it will be severely stung by an insect that looks like the species it attempted to capture as prey. In **warning coloration,** on the other hand, the bright colors and markings of the individual prey species simply advertise its noxious qualities to any potential predator (Fig. 3-9). Bumblebees and brightly colored caterpillars, especially those bearing red or yellow patches, are frequently encountered examples of warning coloration.

Another more ecologically intricate interaction between predator and prey is found in **aggressive mimicry.** Here the predator deceives the prey by its coloration and behavior. Ambush bugs (members of the insect order of "true bugs," Hemiptera) that live in flowers look like part of the flower; therefore it is easy for them to attack unsuspecting bees and flies coming for nectar. Some female fireflies imitate the mating flashes of females of other species and then capture and eat the males of the species that come in response to the mimicked flash pattern.

Symbiosis

Symbiosis refers to a long-term interspecific relationship in which two species live together in more or less intimate association. This is not a social system but an ecological association involving some transfer of energy or adaptive benefit. Symbioses are of three general types: **commensalism, mutualism,** and **parasitism.**

Commensalism occurs when one species benefits from the association but the other species is not significantly affected. Cockroaches benefit quite a bit from living in close proximity to man but humans are rarely harmed or benefited by having them in their homes. They may be aesthetically displeasing commensals to their human hosts, but they hardly eat enough to affect man's energy intake. In the American tropics and subtropical Florida, Old World cattle egrets have become

A

Fig. 3-9. Warning coloration: (A) a bright red beetle (*Tetraopes*) on a milkweed leaf; (B) a bright yellow and black striped moth larva with irritating hairs; and (C) a brilliant red "poison arrow" frog (*Dendrobates*) of the Costa Rican rain forest. South American Indians boil the skins of frogs of this genus to produce a deadly poison.

B

C

well established during the last two decades. These birds feed in pastures very close to cattle, eating the insects that the cows scare up while grazing. As far as is known, the cattle are not benefited or harmed (e.g., by diseases the birds might carry from herd to herd) by the presence of the egrets. However, in the African tropics to which the cattle egret is native, it probably serves as a warning sentry for the grazing mammals, alerting them to possible predator danger by sudden flight. A more intimate commensal-host relationship is that of epiphytic plants, such as arboreal orchids and bromeliads, with the trees on whose limbs and trunk surfaces they grow. These so-called "air plants" gain support (a physical substrate on which to grow) above the ground in a favorable light level, and spend their entire life spans there. But they neither benefit nor harm their host, unless the weight and wind resistance of the epiphytes become so great that the structure of the tree is weakened.

In mutualism both species in the relationship benefit. Lichens, for instance, are mutualistic associations between one-celled photosynthetic algae and chlorophyll-lacking fungi. The algae, through photosynthesis, provide sugars as food for both species while the fungal hyphae (thread-like bodies) collect and hold moisture and mineral nutrients that both symbiotic partners can use. Pollination of flowering plants by insects is perhaps the commonest example of mutualism (Fig. 3-10). Bees depend on flowers for food in the form of nectar and pollen, and the flowering plants depend on bees or other pollinators to carry their male reproductive cells (sperm in the pollen grains) specifically to the female parts of other flowers of the same species. In Africa there is a small mammal species called the honey badger or ratel that has a mutualistic relationship with birds called honey-guides. The honey-guide will fly along in front of the ratel to lead it to a bee nest that the bird has discovered, and after the ratel has torn open the nest for the honey, the bird will feed on the wax of the honeycomb.

Domesticated plants and animals usually enjoy a mutualistic relationship with man. Man cares for them and insures their continued reproduction (without which they would soon die in competition with wild species) while he depends on them for his own existence. In their original wild state, these plants and animals were perfectly capable of survival, but artificial selection by man during the course of domestication has bred out many of these adaptive traits in exchange, as it were, for traits yielding higher food production. The three great prehistoric centers of domestication have yielded most of our commonly cultivated species during the last ten thousand years. From southwest Asia we obtained wheat, cattle, sheep, goats, barley, and rye; from southeast Asia rice, pigs, and chickens; and from our New World tropics we have corn, beans, potatoes, squash, and turkeys, as well as other crops of economic value such as tobacco. In Africa today considerable attention is being directed toward the possibility of domesticating some of the larger antelope species to breed on a large scale for meat production. This possibility is of particular import because essentially we have not added any new species to our assemblage of domesticated plants and animals for the last several thousand years. Thousands of years ago, men everywhere subsisted by hunting game animals and gathering wild plant products. Today man really only predates significantly on fish and a few other wild species in the oceans, and

A

Fig. 3-10. (A) A checkerspot butterfly (*Euphydryas anicia*) pollinates Colorado sunflowers during its visits for nectar, an example of mutualism; (B) A purple sunbird (*Nectarinia asiatica*) pollinates the flowers of *Capparis aphylla* in India while it sips nectar, a less familiar example of a mutualistic interspecific relationship.

B

modern agricultural techniques have made gathering (of blueberries, for example) more a novelty than a necessity. The domestication of certain plants and animals in a mutualistic association with man allowed man to break the continuous cycle of hunting and gathering that he had previously been forced to follow. Hence we can truly say that our large cities and advanced civilization today were made possible by this rather recently established mutualism.

In parasitism one species, the **parasite,** is benefited while the other species, the **host,** is harmed. Parasitism, like predation, is frequently an important factor in controlling natural populations. But despite the similarity of function and definition there are significant differences between parasitism and predation. Parasites are much smaller than their hosts, living as they do on a portion of the host's energy intake. A predator is almost always larger than its prey. Parasites live on, in, or near the host (Fig. 3-11), whereas predators associate closely with the prey only when actually feeding on them. Parasites will kill their host only very slowly or not at all;

Fig. 3-11. Cocoons of a parasitic braconid wasp on a sphinx moth larva (*Manduca quinquemaculatus*).

predators kill their prey quickly and consume it immediately. The most efficient parasite is one whose presence is hardly noticeable to the host; should it kill its host before parasite reproduction the parasite would die without leaving any offspring and a species with this habit would rapidly become extinct.

Ectoparasites live on the body surface of the host and possess modifications for secure attachment to the host species. Examples include ticks, lice, fleas, and bedbugs. **Endoparasites** live inside the host, frequently in the digestive tract. Consequently they exhibit adaptations for resisting the internal movements and defensive measures of the host's body. For example, tapeworms and flukes must have outer skins that will protect them against digestive secretions and cellular enzymes. The life cycle of endoparasites frequently includes intermediate hosts between the egg and adult parasite stage, as in the Chinese liver fluke life cycle, where snails and fish host larval fluke stages (Fig. 3-12). The fluke larva is passed to man when he eats an intermediate fish host. In man the fluke becomes an adult and produces great quantities of eggs, which pass out of the human body in the feces. The eggs hatch and the new larvae infect snails, starting the cycle again. Later larvae leave the aquatic snails to infect fish. Occasionally in nature one encounters the phenomenon of **hyperparasitism,** where one parasite will live in another parasite which in turn is living in a host. An example would be a wasp larva parasitizing a fly larva that is parasitizing a moth caterpillar.

These three basic types of symbiosis—commensalism, mutualism, and parasitism—are all profoundly important interactions in biological communities. Without the mutualistic association between nitrogen-fixing bacteria and legume plants, for instance, the nitrogen biogeochemical cycle would be far slower and would no doubt have limited the expansion and evolution of life on earth to a lower level than we have today. In company with predation, certain of these symbioses exercise important density-limiting functions on populations of other species. Let us now look at another complicated set of interactions within and between species; we may term it competition.

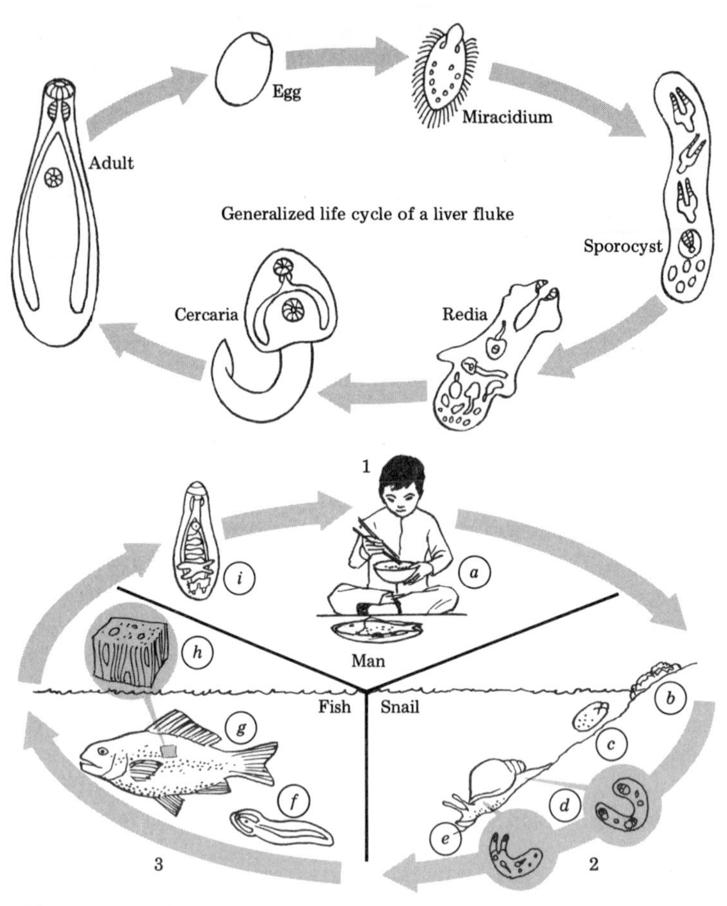

Fig. 3-12. The extreme complexity of some parasitic relationships is exemplified by the life cycle of the Chinese liver fluke (*Clonorchis sinensis*). The life cycle involves more than one host (man, snail, and fish), and there is much asexual budding, which increases the total reproductive potential. *Top figure:* the succession of developmental stages in the life cycle. The adult releases eggs which hatch as miracidium larvae. This larval type becomes a sporocyst which buds, internally, a large number of redia larvae. These new larvae in turn bud, internally, a large number of cercariae, tadpole-like larvae which develop into adult liver flukes. *Bottom figure,* the succession of hosts. *1.* Man: (*a*) Eggs are liberated in human feces (*b*) reaching fresh-water ponds, where the miracidium larvae (*c*) swim about and, on contacting a snail, enter it. *2.* Snail:

Within the snail the miracidium develops into a sporocyst (*d*) that, in turn, liberates rediae (*e*). The rediae enter the snail's liver; here they themselves reproduce, and some of these eventually produce cercariae (*f*). *3.* Fish: The cercariae leave the snail and swim free in the pond, eventually entering a fish in whose muscles they encyst. The infective cycle is completed when the fish is eaten by man; the cysts germinate and the adult fluke parasitizes man.

Competition

Competition in an ecological sense is a struggle between organisms for food, space, mates, or any other limited resource. In intraspecific competiton the struggle is between individuals of the same species. This experience is common to all species whose numbers are increasing in a limited environment; in a later chapter we will treat it as one of the central factors regulating population growth. In interspecific competition the struggle is between the populations of two species. The basis for such struggles has been expressed as **Gause's Principle,** which states that two species cannot occupy the same niche simultaneously. If two species that occupy the same niche (way of life) come together in space and time, as a general rule there are three possible results.

Extinction

In this outcome, one species becomes **extinct** (locally) because its competitor was more successful at monopolizing available resources such as food or space. G. F. Gause formulated his principle after studying competition between two species of *Paramecium,* tiny one-celled aquatic animals in the phylum Protozoa. These can be grown in laboratory culture vessels of limited size where the *Paramecia* compete for food (bacteria), oxygen, and other requirements. They reproduce themselves by splitting in two (binary fission) as often as several times a day. When grown alone in containers in the laboratory, the numbers of *Paramecium caudatum* increase more slowly than *Paramecium aurelia* (Fig. 3-13). Eventually each species reaches the capacity of that environment to support the population, and growth levels off to a population size characteristic of the species. When grown together in the

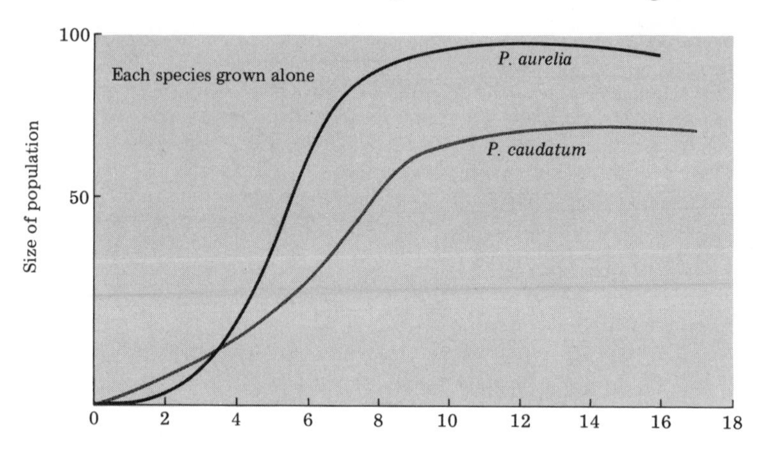

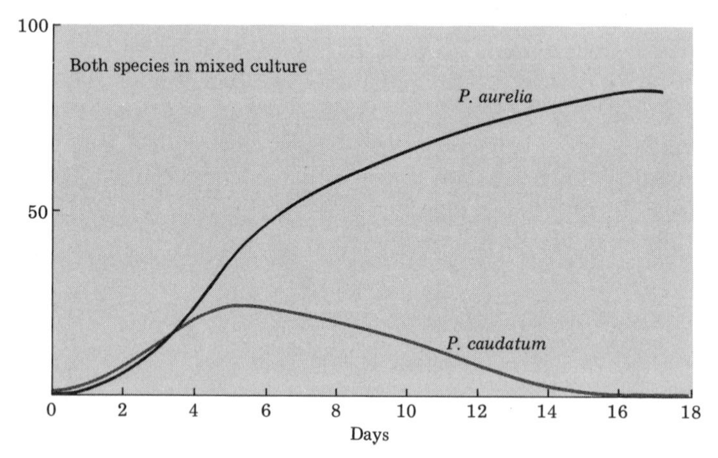

Fig. 3-13. (*Top*) The growth of each of two species of closely related *Paramecium* when grown alone. *Paramecium aurelia* and *P. caudatum* exhibit normal sigmoid growth in controlled cultures with a constant food supply. (*Bottom*) Effect of competition between *Paramecium aurelia* and *P. caudatum* when grown together; the latter species is eliminated.

same containers, *P. aurelia* increases more slowly than it did alone and *P. caudatum* grows poorly and eventually becomes extinct (Fig. 3-13, bottom). Three facts may be noted in this competitive situation: (a) normal growth rate was experienced by each species when first combined, as long as food was not limited in relation to the size of the population; (b) the population of *P. caudatum* declined to extinction eventually because of competition from *P. aurelia;* (c) when food became limited and competition was acute, *P. aurelia* grew more slowly than when cultured alone, but it gradually attained the normal population size. *P. aurelia* was the more successful competitor because it was adapted to rapid growth and hence could trap a greater percentage of the limited available energy (bacterial food) in each generation.

Competitive Exclusion

In **competitive exclusion** one species is forced out of part of the habitat when it comes together with its competitor, but it continues to survive in other adjacent portions of the habitat. An experimentally verified case of competitive exclusion is seen in the vertical distribution of two species of barnacles along the coast of Scotland. *Balanus* barnacles normally live on coastal rocks below the mean high mark of the lowest tide each lunar month; thus these barnacles are always covered with water for part of each twenty-four hour period. *Chthamalus* barnacles live on rocks in the splash zone above the mean high neap tide mark (Fig. 3-14). This distribution suggested competitive exclusion as an explanation, but experiments had to be made to eliminate the possibility that the species are restricted in their distribution not by competition but by inability to adapt to other parts of the shoreline habitat. So the biologist J. H. Connell removed *Chthamalus* barnacles from splash zone rocks and found that the *Balanus* species still would not live there. When he removed *Balanus* from underwater rocks, however, he found that *Chthamalus* did move down and survive below the water line. Thus competitive exclusion is involved in limiting the distribution of *Chthamalus* wherever both of these species occur together; the young *Balanus* simply grow more rapidly than the young *Chthamalus* in the lower zone and vigorously over-grow or pry off their competitors.

Character Displacement

Where two potentially competing species occur **sympatri-**

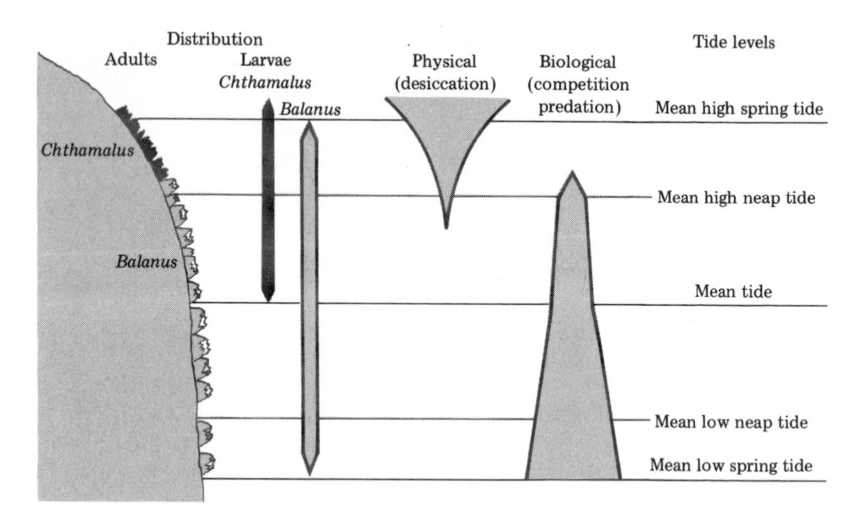

Fig. 3-14. Competitive exclusion effect of competition between two species of barnacles. Although there is a broad area in which the larvae of both species settle, competition eliminates most of the overlap by the time the adult stage is reached, *Chthamalus* being largely restricted to the zone above the level of the mean high neap tide (neap tides are the lowest tides of a lunar month) and *Balanus* to the zone below this level.

cally (together), they have greater differences in their feeding adaptations (such as bill size in birds) than in areas where they do not coexist. Each species has evolved characteristics that lessen competition with other species for food resources and hence promote reproductive success. This divergence of characters in areas of overlap is called **character displacement.** In birds, for instance, bill sizes will differ distinctly between related sympatric species, and one species with a long thin bill will feed on large insects while another with a short thick bill will feed on seeds. These differences minimize the competition that would result if both species were similar enough to feed on the same prey. Where these bird species occur alone (**allopatrically**), the beak size of each will often be similar, adapted for generalized feeding on a variety of food items. A hypothetical example of character divergence, this adaptive consequence of the influence of competition in areas of sympatry, is shown graphically in Figure 3-15.

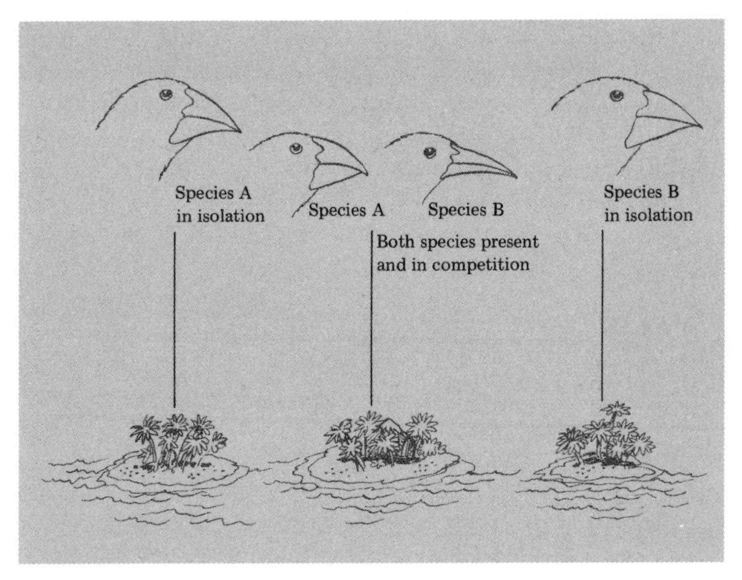

Fig. 3-15. Character displacement effect of competition between two hypothetical species of finches on an island chain. Where species A or species B is found alone on an island, the beaks are generalized in shape and hence are adapted for feeding on a whole variety of food sources. A local evolutionary change in beak size occurs when both species are found on the same island, allowing each species to specialize on different food resources.

Summary

Three basic types of interspecific relations in a biotic community are predation, symbiosis, and competition. The predator and prey populations in a long-established predation relationship are usually in a dynamic balance that benefits both species. A common evolutionary result of predation pressure is protective coloration in the prey species. Symbioses involve long-term intimate association with some transfer of energy or adaptive benefit. In commensalism, one species (commensal) benefits while the other (host) is not affected by the association. Both species benefit in mutualism. In parasitism one species (parasite) benefits and the other (host) is harmed. Competition represents a mutual struggle, either intraspecific or interspecific, for a limited resource. Gause's Principle states that two species cannot occupy the same niche simultaneously. Hence interspecific competition usually results in extinction of one species, competitive exclusion, or character displacement.

4
THE ORGANIZATION OF POPULATIONS

Properties of a Population

Populations are like the individual organisms that constitute them in that they are living, have a definite structure and orderly functioning, and grow and die. Many of the principles of ecology that we have considered in the preceding chapters apply to the population level of organization as much as to communities or ecosystems, and we shall begin our consideration of population biology in this chapter with an overview of the organization of these biological units.

Before continuing, though, let us first define what we are talking about: a **population** is a group of organisms of the same species living in a particular space. We cannot be much more specific than that, for populations vary greatly in size and character depending upon the species of organism involved and the limits of the space it occupies. For example, we frequently consider the species of man (*Homo sapiens*) to be a population over the whole earth or a population as small as that of a tiny mountain village; either application of the term "population" is technically correct. Of course, until the development of modern transportation, people on different continents and even people separated by natural geographic barriers on any one continent composed reasonably isolated populations.

Some of the properties of a population are unique to the aggregation and are not characteristic of the individuals in the group. Under the general category of organization, a population can show **dispersion** (distribution), **dispersal** (moving out from a center of origin), and **density** (degree of crowding), while the growth of a population includes its **natality** (birth rate), **mortality** (death rate), **age distribution** (percentage of the population in each age class), **biotic potential** (maximum possible growth rate under ideal conditions), and **growth form** (the manner and speed of population growth). Each of these properties influences the ecological status and evolutionary future of a population, and when we understand their normal role it becomes easier to see how environmental pollution and other changes wrought by man can radically affect natural populations of organisms. It also clarifies the role of similar properties in the population biology of man himself. Let us look at these particular population parameters in more detail now.

Dispersion

Dispersion is simply the internal distribution pattern of individuals within a population. Organisms in a population may be distributed according to three general patterns: **random, uniform,** or **clumped** (Fig. 4-1).

Random Dispersion

Random distribution—individuals scattered over an area without any regularity or any degree of affinity for each other—is relatively rare in nature. It occurs only where the environment is very uniform, with resources spread evenly throughout the area of the population's distribution, and only when the species has neither a tendency to aggregate because of social attractions nor a tendency to repel one another. Since we usually find a non-random dispersion of resources in nature, organisms are almost never random in distribution. Solitary predatory spiders living on a forest floor uniformly

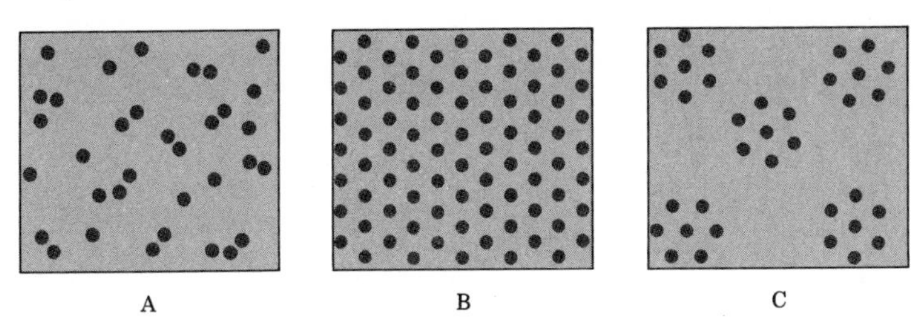

Fig. 4-1. The three basic patterns of dispersion of individuals in a population: (A) random, (B) uniform, and (C) clumped.

covered with similar leaf litter and with uniform moisture conditions will, however, show random dispersion.

Uniform Dispersion

This type of distribution is more regular than random dispersion, yet no aggregation of individuals occurs. Uniform distribution is not infrequent in nature and may occur where competition between individuals for resources is severe or where positive antagonism promotes even spacing. This type of distribution occurs in island nesting colonies of sea birds, where nests are often spaced apart by the length of the mother birds' neck-to-bill-tip reach! But it is more common among plants than animals. The creosote bush and other desert plants of southwestern North America compete intensely for the little annual moisture that falls. The roots of many species actually secrete antibiotics that inhibit the growth of seeds or neighboring plants within the radius of their root systems.

Clumped Dispersion

Clumped dispersion is defined almost by elimination: it is irregular, non-random distribution of individuals in the population. Clumping of varying degrees represents by far the most common pattern of dispersion of organisms in a population area. This clumping distribution may vary in intensity and even be present only at certain times of the year or of the organism's lifetime. The clumps or aggregates themselves may be scattered uniformly or at random throughout the area of distribution, depending upon the distribution of resources or degree of social interaction between the sub-groupings within the population.

The reasons why clumping is common are not hard to understand. Individuals respond to local habitat differences within the population's area of distribution, seeking out or surviving optimally in habitats with the best combination of environmental factors (temperature, light, minerals, water, etc.), which are rarely distributed uniformly. At certain seasons of the year weather conditions may promote aggregation. During the dry seasons, populations of most wildlife species in Everglades National Park in south Florida will aggregate around "gator holes," depressions in watercourses that have been made by alligators, where water will stand long after the open flat marshlands have dried up. The degree of sociality of a higher animal species is important in determining aggregations within a population. Many mammals and birds feed and function best as groups, and hence social attraction promoting clumping is adaptively advantageous to them. Reproductive processes require at least one male and female cohort to aggregate, and depending upon the type of reproductive pattern characteristic of the species, the grouping may involve any number from a pair of organisms to a whole colony of breeding individuals.

The above points are some of the obvious reasons for animal and plant aggregation, but what of the ultimate evolutionary reasons for most clumping behavior? In a positive sense, we can observe that aggregation often increases the *survival* of the individuals in the group. Individuals found in clumps tend to have a lower mortality rate than isolated individuals. The group offers *protection* against attacks by predators, and will frequently *modify* the microclimate or microhabitat where they live in such a way that unfavorable climatic periods do not affect the grouped individuals as much as they do solitary organisms. Sufficient heat is generated within a beehive to allow the colony to survive cold night temperatures that would kill any isolated bees. The chief negative aspect of aggregation is that it almost inevitably *increases competition* for light, nutrients, food, or space.

Social Organization and Dispersion

As we have noted, the degree of sociality of an animal species will affect the distribution of the members of a population. Three principal types of social organization that affect dispersion are **territoriality, social hierarchies,** and the existence of actual **societies** in which all individuals of a population live.

Territoriality

Most vertebrates and higher invertebrates commonly live their whole lives in a definite area. This area of normal activities is called a **home range.** This is where the animal searches for food, searches for mates, rears its young, and so forth. If all or part of this area is actively defended, the defended portion is termed a **territory.** An individual (or a pair or a family group) will defend such an area against intrusion by any other individual. This naturally results in a *spacing out* of the population, frequently in direct relation to the resources (especially food supply) of the environment other than mere physical space. We can see a nice example of this effect of territorial behavior in Howler Monkey (*Alouatta villosa*) populations inhabiting the lowland forests of the New World tropics. A small population group of these monkeys travels around as a unit, maintaining a "core" of territory against all other Howler groups. It does this by communal howling in the early morning hours, which warns other groups of Howlers of their presence near a defended territory (Fig. 4-2).

The advantages of the dispersion resulting from territorial isolation are threefold: (1) interindividual competition is reduced; (2) energy that would be spent in antagonistic fighting if all individuals were close together is conserved during critical periods such as the mating season; and (3) overcrowding and exhaustion of the food supply is prevented.

Fig. 4-2. Spacing of individuals by group territories, in this case territories of three troops of Howler Monkeys in the tropical deciduous forest of Costa Rica. The home ranges of the three bands actually overlap; only the "core" territory areas are shown. Dots represent the howler monkeys in each troop. At dawn all the males howl vociferously for about an hour (see inset) and each troop can determine the position of the neighboring groups without actual physical contact or fighting. The result is a well-spaced series of populations that do not exhaust the resources of the environment.

Social Hierarchies

A social hierarchy is simply a series of dominance-subordination relationships in a population. You may have heard this concept referred to as a "pecking order," since step-like series of dominance relationships are commonly observed among chickens and other barnyard fowl. The most aggressive individual is at the top of the pecking order and is free to peck all the other birds. The least aggressive individual is at the bottom and may be pecked by all the birds above him in the social hierarchy. In a closed system like a flock of ducks on a farm, it may be hard to see how this social relationship could extend to dispersion. But in the wild, such hierarchies frequently result in the expulsion of a certain subordinate segment of the population, which is forced to move to new ranges. In African baboon troops, small sub-dominant groups containing both males and females split off from the main troops when these latter units run by dominant adult males grow quite large (Fig. 4-3). In deer, elk, or antelope herds at rutting season, a dominance hierarchy of males is established, with the most powerful males each obtaining control over their own harem of females. Juvenile and mature subordinate males are forced by the dominant males to disperse into adjacent areas of the population's home range. After the mating season, the herd reforms.

Societies

Sometimes organisms form highly organized societies in nature, outside of which an individual cannot survive. It may seem confusing to define a population for species where a colony or society contains one reproductive pair and all the other individuals are sterile workers. The social insects, such as termites, ants, wasps, and bees, are good examples of society-dependent organisms where a "population" of the species may properly refer to all colonies within a particular region. The colonies are spaced out in the environment according to the availability of resources necessary to maintain them. Each colony is actually a genetic extension of one reproductive pair; that is, all the non-reproductive workers and other castes arise from this pair's genetic contribution through the queen's production of millions of eggs during her lifetime. As many as several million sterile individuals may work together to maintain a reproductive pair in a colony. Because new colonies are founded by specially produced reproductive adults leaving the mother colony several times a year and settling nearby, a population's dispersion really includes all the colonies in such an area. Colonies that are too close to each other risk extinction as a result of competition for food. Hence in an adaptive sense there is strong selection for an adequate dispersion of colonies. The emerging new queens have to fly a sufficient distance away from their "home" colony or the society they will found by egg-laying will soon be lost through competition from the older, larger, and better-established colonies in the overall population.

Fig. 4-3. The positions of group members during movement of a baboon troop at Nairobi Park, Kenya. Dominant adult males accompany females with small infants and a group of older infants in the group's center. A group of young juveniles is shown below the center and older juveniles above. Other adult males and females precede and follow the group's center. Two estrous females (dark hindquarters) are in consort with adult males. New troops are probably formed by splitting off a sub-dominant groups when the original troop becomes quite large.

Population Dispersal

All our discussion on dispersion (the pattern of distribution of organisms *within* a population) has assumed normal individual movement on a local scale. Now we may consider the problem of population dispersal in a wider sense, which is the movement of individuals into or out of the population area. This takes three forms: **immigration, emigration,** and **migration.**

Immigration is one-way movement into an established population area or an uninhabited area from neighboring areas. This is normally successful for the immigrants only when the population into which they are merging is undercrowded in relation to its available resources. *Emigration* is one-way movement out of a particular population, and commonly occurs when overcrowding results from excessive reproduction or an environmental stress such as drought. *Migration* is the periodic departure and return of individuals to and from a population area. We are often filled with wonder and excited intellectually at the remarkable feats of animal migration recorded in the annals of natural history and ecology. It is well worth considering in detail several of these migratory scenarios, where careful biological investigation has indicated the functioning of emigration and migration as dispersal factors shaping population growth and density.

The green turtle, *Chelonia mydas,* is found in oceans around the world and originally became famous because of its use by London soup chefs and for victualing ships. Populations are genetically isolated by their habit of mating and egg-laying at widely separated sites, even though much of the intervening time is spent on communal feeding areas. One such population lays its eggs on the beaches of remote Ascension Island in the middle South Atlantic, located over 1200 miles from the eastern coast of South America (Fig. 4-4). Yet in the extraordinary life history of the turtle, this site is visited by any one female only at two- or three-year intervals. During the many intervening months she feeds with the other adults along the shores of Brazil. When it is time for her to

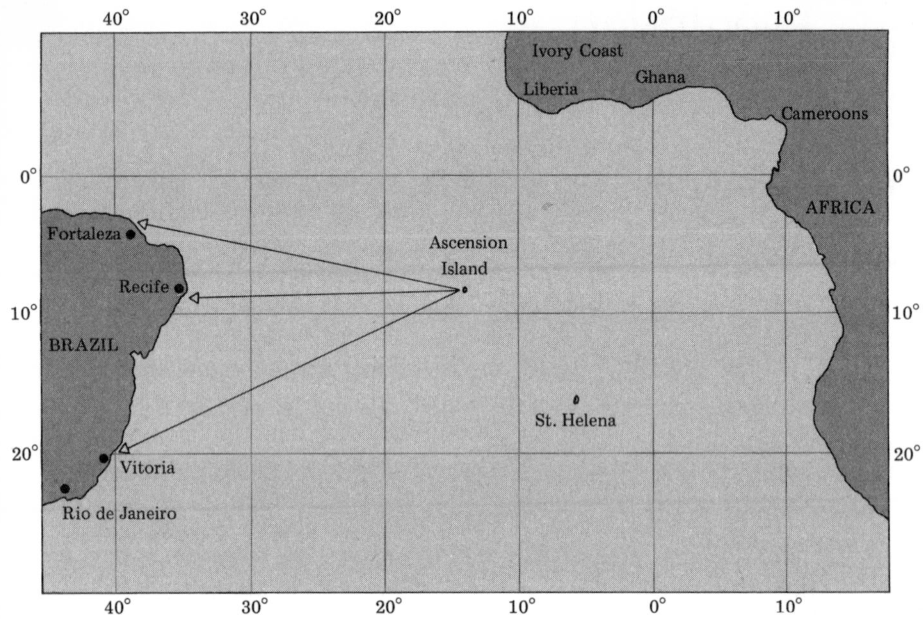

Fig. 4-4. Map of the middle South Atlantic, showing the position of remote Ascension Island, where green turtles (*Chelonia mydas*) nest, and the eastern coast of Brazil, where the turtles disperse on feeding grounds and spend most of their lives.

lay her eggs, she swims 1200 to 1400 miles from the mainland to the tiny (six by seven miles) oceanic island, apparently navigating by celestial or olfactory clues on the high-seas portion of the journey. Appearing at the Ascension Island nesting grounds in April with hundreds of other females and males from the Brazilian coast, she makes her way ashore. She lays about 115 eggs in a pit she digs in the sand and then leaves the beach, not to return for another several years (Fig. 4-5). During this visit she mates with a male in the offshore surf, fertilizing the developing eggs that will be laid two or three years hence. When the eggs hatch about 60 days after oviposition, the tiny hatchling turtles rapidly navigate across the beach to the sea and swim off into the Equatorial Current, which they apparently follow downstream to the bulge of Brazil. There they grow up on a diet of turtle grass in the rich offshore feeding grounds, and repeat the cycle of return with the adults to Ascension Island when they mature some years later. When they return, over 80 percent of them will "home" to the exact island beach on which they had emerged.

To explain this remarkable sequence of migratory behavior one must emphasize its ecological benefits. By utilizing the warm shallow waters of the Brazilian coast and Caribbean for feeding purposes, the sizable population of thousands of green turtles that must come together at the time of reproduction is able to disperse over a broad area and thus reduce grazing pressure on the turtle grass. If all these turtles remained permanent residents of a restricted area along the Brazilian coast, nesting and feeding in the same site, probably only a small population could be supported by the marine environment. By nesting briefly on remote beaches the species is protected from predation pressure on both the eggs and young turtles. This sudden influx of possible prey on the beaches is too brief and too infrequent to allow predators to develop evolutionary specialization on the vulnerable stages

Fig. 4-5. Female green turtle nesting at Ascension Island. Digging has been completed and laying begun. The more pronounced and more angular emargination of the shell over the neck of the Ascension turtle is typical of this population.

in the turtles' life cycle. In sum, periodic migratory behavior and the cycle of adult concentration (breeding) and dispersal (feeding) seems to increase population stability and hence survival of the green turtle.

In the arctic lemmings, however, emigration is a response to overcrowding and no return movement occurs. Periodically these small rodents increase tremendously in numbers, and in northern Europe they have been known to march by the millions across the land, devouring everything edible in their path. Upon reaching the coast, many die by drowning. Predators become so satiated by the immense numbers of prey that they are ineffective in controlling these outbreaks. It is only the migratory behavior that accomplishes the population reduction necessary to bring the lemming numbers down to the level that the environment can support. We shall consider population regulation in these interesting mammals in more detail in the next chapter. Meanwhile, we may merely note that the ecological advantage of population regulation through emigration is that the remnant of animals left behind will be able to continue to breed and form the nucleus of the next generation, without the fatal environmental stress of continued overpopulation.

Migratory waterfowl are classic examples of organisms that utilize long-distance movement as an adaptive response to seasonal abundance of food. In the late spring of the year these ducks and geese fly north to the arctic, where they mate, nest, and raise their young. When these young are able to fly in late summer, the adults and young leave the nesting grounds together and fly several thousand miles south to the tropics where they spend the winter.

As a result of studies of bird-banding data on ducks and geese, Frederick Lincoln, a biologist with the U.S. Fish and Wildlife Service, discovered and reported in 1935 the existence of four great migratory flyway systems in North America. Later work has shown that most if not all of the other American migratory birds follow the same routes. Each flyway is a vast geographic region with extensive breeding grounds and wintering grounds connected with each other by a system of migratory pathways. A particular flyway has its own populations of birds, which use the same route every fall and spring. During the nesting season, the northern breeding grounds are shared in many areas by birds of the same species but belonging to different flyways. The extent of the Pacific Flyway across the North American continent is illustrated in Figure 4-6.

The tropical wintering grounds provide adequate food for the migratory adults as long as they are there only part of the year. Competition with resident species for food and space would, however, limit a waterfowl population that bred in the tropics. Thus during the breeding season these migratory birds take advantage of the several months' abundance of food in the arctic and the ample territorial space there in which to rear their young.

Of course, beyond social organization and various evolutionary drives for adaptive migratory behavior, we must recognize that the dispersal of an organism is greatly influenced by natural barriers. These may be ecological barriers, such as unsuitable habitats, or broader geographic barriers like mountains to a desert species or a large body of water to a strictly terrestrial species. Also, the **vagility** (inherent ability for movement) of an organism will greatly affect dispersal. The capacity of most birds to fly long distances greatly exceeds that of most butterflies, for example. In all of this discussion, we must remember that the movement of the **disseminules** * of individuals, such as seeds, eggs, larvae, and spores, are just as significant ecologically in establishing the dispersion of a species as the movement of the mature adults themselves. Ultimately, the ecological meaning of dispersion and dispersal is found in the growth and regulation of the population unit itself.

Summary

A population is a group of organisms of the same species living in a particular space. Such an aggregation possesses prop-

* Disseminules are any stage of the life cycle of an organism which is capable of being dispersed, but this term usually refers to the immature stages.

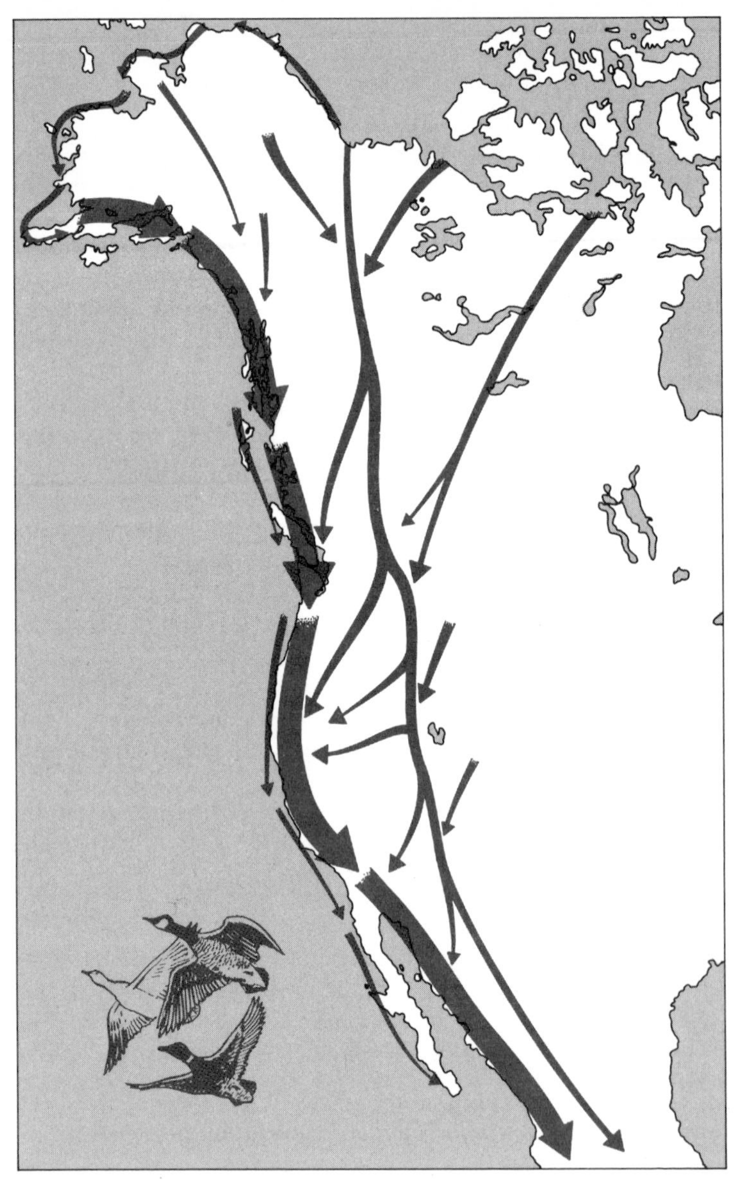

Fig. 4-6. The Pacific Flyway for migratory waterfowl in North America.

erties unique to the group, such as dispersion, dispersal, density, natality, mortality, age distribution, biotic potential, and growth form. In this chapter we considered the first three characteristics under the category of population organization. Dispersion, which is the internal distribution pattern of individuals within a population, may be random, uniform, or clumped. Clumping, the most common type, is caused by non-random local habitat differences, social attraction, and reproductive processes, although aggregation almost inevitably increases competition between the individuals involved. Territoriality, social hierarchies, and colonial species or societies are principal types of social organization that affect dispersion. Movement into or out of the population area involves three kinds of dispersal behavior: immigration, emigration, or migration. Natural barriers, adaptive behavior, and the vagility of the species greatly affect dispersal.

5 POPULATION GROWTH AND REGULATION

For everything there is a season,
and a time for every matter under heaven:
a time to be born, and a time to die;
a time to plant, and a time to pluck up what is planted;
a time to kill, and a time to heal;
a time to break down, and a time to build up . . .

Ecclesiastes 3: 1–3.

Population growth is the increase in number of individuals comprising an aggregation. As we shall see shortly, it is not necessarily the result of more births than deaths, but may be caused by increased survivorship, movement into the area of new organisms of the species under consideration, or other factors. Growth of a population without increase in emigration or removal by other means does cause an increase in **density,** which is simply the size of the population within a particular unit of space.

Density

Population density is usually measured and expressed as the number of individuals per unit area (or volume). We may measure the **crude density,** that is, the number (or biomass) per unit of *total* space, including all environment within the described boundaries. Alternatively, the ecologist may measure the number (or biomass) of organisms of a species per unit of *habitat* space. That is, the area that is actually suitable for a species to live in may be much smaller than the total space, which might include rocks, ponds, or other unsuitable habitats. In this case the measurement is known as the **specific** or **ecological density.** It states the number of individuals in the area or volume actually available to be colonized by the population. Naturally, there is always some amount of space within a population's dispersion that simply cannot be occupied because it is used by other organisms or because of physical features of the environment. Thus the specific density is a preferable measurement.

A common difficulty is that one can measure this density (crude or specific) at a time when the population is changing in size, and the estimate may not mean too much. Especially in the lower plants and animals, such as algae and soil insects, with short generation times, population size can change *very* rapidly. Under such conditions many means of population measurement will give grossly underestimated or overestimated figures. Accurate measurement may also be very difficult because of the large area the population is inhabiting or the irregular distribution of individuals. Thus ecologists frequently find an **index of relative abundance** a more useful statistical reference than an absolute density figure. To obtain it, rather than counting the number of mealybugs per acre in two different groves, for example, one counts the average number of mealybugs per orange tree in each area. The statement that there are 5,000 mealybugs per orange tree in grove A versus 20 mealybugs per tree in grove B is a much better indicator of the relative severity of infestation than calculating a "guesstimate" of the numbers per acre, when tree hosts may be of different ages and sizes in the two areas or more widely spaced in one than the other.

Of course, whichever measurement of density is used, it must be recognized that populations are rarely static in size and that there is usually an input of new individuals into a population by births and immigrants as well as a loss by deaths and emigrants. The resulting trend in numbers we call positive or negative **population growth,** depending on whether input exceeds loss, and it inevitably influences density. The component factors involved in population growth will now be examined, and then we shall take up the subject of the regulation of population growth and size.

Growth

Involved in the overall growth of a population are three major factors: natality, mortality, and survivorship. These determine the pattern of increase (or decrease) for a population.

Natality

The natality or **birth rate** of a population is simply the increase in the population due only to the normal rate of reproduction. The maximum possible natality, of course, is rarely reached because the ideal conditions that would be the prerequisite of maximum production of new individuals do not exist in natural situations. Ideal conditions would exert no ecologically limiting influences, and thus reproduction would be limited only by physiological factors, such as the maximum egg production possible per unit of time. Therefore we will normally use "natality" to refer to the observed population increase from reproduction under a particular set of ecological conditions. Unlike the maximum possible natality for a species adapted to a certain area, this actual birth rate is not a constant, but will vary with the size and composition of the population as well as physical environmental factors.

Natality is usually expressed in mathematical notation as

$$\frac{\Delta N}{\Delta t} \quad \text{or} \quad \frac{\text{number of new individuals produced in population}}{\text{change in time}}\,^{*}$$

This indicates the production of new individuals per unit of time. The natality rate for a flock of hens producing 90 new chicks per month would be approximately 3/day (90

* The Greek letter *delta* (Δ) in mathematical notation means "change in" the quantity following the delta.

chicks ÷ 30 days). Demographers, specialists in the study of human populations, usually express natality or birth rate as the number of live births per one thousand members of the population per unit of time. Here one divides the total number of new babies for the year by the size of the population at the midpoint of the stated time period to find the number of births per person. One then multiplies this figure by 1,000 to get the number of births per thousand persons. Thus, during 1971 there were approximately 3,769,000 live births in the United States. At mid-1971, the country's population was estimated to be 207,100,000. Therefore the birth rate for the year 1971 was 3,769,000/207,100,000 = 0.0182. Since this is the rate per person in the population, we multiply 0.0182 by 1,000 and see that the United States birth rate was 18.2.

Mortality

The mortality or **death rate** of a population refers to the number of individuals dying per unit of time. The minimum possible mortality equals the population loss under ideal or nonlimiting conditions. That is, even under the best environmental conditions, individuals will die of old age, at a point determined by their **physiological longevity.** The actual observed mortality is the rate of loss of individuals under given environmental conditions where predators, accidents, competition, and other factors share in causing deaths. This actual mortality is what we normally consider the mortality rate of a natural population.

While the death rate for most species is expressed as number of organisms dying per unit of time, e.g., 85 deer dying per year in the population inhabiting a particular forest, demographers describe mortality in human populations as the number dying per one thousand members of the population per year. In 1970, for example, the death rate for the United States was 9.6 people per thousand individuals in the total population.

Survivorship

The percentage of individuals living at various ages is the **survivorship** in a population. This is usually expressed in a survivorship curve: the surviving percentage per thousand born is plotted from the maximum of 1,000 alive at birth to the death of the last individual at the maximum life span for the species. Sample survivorship curves representing four types of situations are illustrated in Figure 5-1.

In a population of starving adult fruit flies in a laboratory bottle (curve A), most of the *Drosophila* live out the potential life span. After this long period of a nearly horizontal survivorship curve, all the flies begin dying at once, causing a precipitous drop in survivorship near the point of maximum life span under starvation conditions. This survivorship pat-

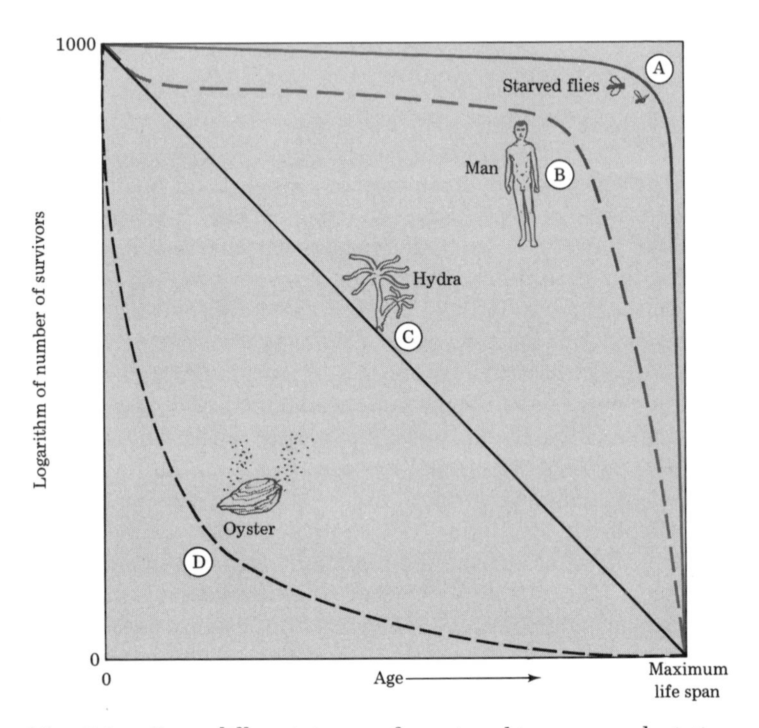

Fig. 5-1. Four different types of survivorship curves, depicting the number of individuals still living at various times after birth. (A) Survivorship of a population of starved adult fruit flies starting with emergence from pupae. (B) Survivorship of a population of man. (C) Survivorship of a *Hydra* population. (D) Survivorship of a group of 1,000 oysters from the time of hatching of the larvae to end of adult life span.

tern is very rare in nature since environmental factors normally kill some fraction of a population early in life.

The situation represented by curve B, with an initial drop in survivors and then a period of good survival until death of individuals in old age, is representative of man and many other larger animals. If the organisms survive infancy, with its relatively high death rate, they have a good chance to live out almost the entire maximum life span for the species in that region.

When the mortality rate at all ages is constant (curve C), one sees a survivorship curve like that of the population of *Hydra*, a freshwater coelenterate. Fledged birds also show this type of survivorship: mortality at any age after fledging (acquisition of flight feathers) is equally probable.

If, on the other hand, most individuals die early in life and the survivors of that period then have a comparatively low death rate, we have a curve (curve D) similar to that of the oyster. Tremendous numbers of oyster larvae die soon after hatching because of the hazards they meet as they are moved by the currents through their marine environment. These tiny larvae are particularly vulnerable to predation and are very likely to land in places where they cannot survive, when they settle on the bottom to begin life as sessile adults. Thus the survivorship curve for the population drops rapidly at first. The few larvae that attach to a satisfactory substrate will survive well.

Obviously, the pattern of survivorship for a species' population depends on the extent of environmental pressures on each age group. Should all environmental restraints be absent, one would expect a typical survivorship of 100 percent until the maximum life span had been reached. In this case, we would expect to see the maximum possible rate of increase for the population, since all individuals in each generation would survive to reproductive age. Though this rarely if ever happens, the concept of this **biotic potential** is useful because it allows us to estimate the extent of operating ecological restraints (**environmental resistance**).

Biotic Potential

The biotic potential is simply the growth rate inherently possible in a population under ideal conditions. Thus this maximum growth rate is equivalent to the maximum natality less the minimum mortality, with no restraints such as lack of food, predation, parasitism, or competition for space. Since these growth-restraining factors are normally present in natural situations, we actually find and measure only the *realized* rate of increase. The difference between the biotic potential, as measured under ideal conditions in the laboratory or from estimating total egg production, and the realized rate of increase is a measure of the environmental resistance present—that is, all the limiting factors in the environment acting on this particular population.

As a result of the interaction of the biotic potential and environmental resistance, populations tend to have a characteristic pattern of increase or **population growth form,** which we will now now consider in detail.

Population Growth Form

When conditions are nearly ideal, a population may increase very rapidly indeed, at a rate approaching its biotic potential. In 1879–1881 a total of 435 striped bass were released in San Francisco Bay after transport from the Boston area. Striped bass were not native to this Pacific marine environment, and their natural enemies and other limiting factors were apparently absent. By 1899 more than one million pounds of striped bass were being harvested *per year* from San Francisco Bay. This explosive population increase represents one extreme of population growth form, the **J-shaped** or **Exponential Growth Curve.**

J-shaped or Exponential Growth Curve

In species or situations where this type of growth form is

permitted, population density increases rapidly, in exponential or compound interest fashion (Fig. 5-2). The more individuals are added to the population, the faster it increases, because all those that are added also breed and hence increase the total growth rate of the population. The process is similar to bank interest compounded daily, the interest being added to the principal balance and thus providing a larger base on which to calculate the interest for the next time period. This exponential J-shaped growth rate may stop abruptly as environmental resistance becomes effective more

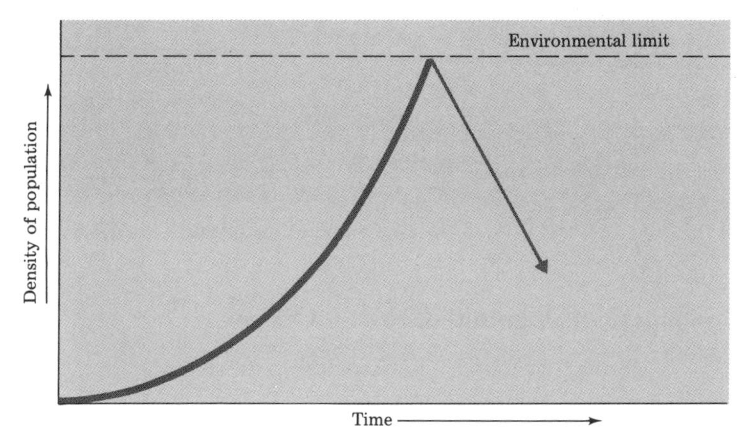

Fig. 5-2. The J-shaped or exponential growth curve of a population. After an initial establishment phase, the population increases exponentially until the environmental limit causes a population crash.

or less suddenly. The population then suffers a crash in numbers, regardless of population density. This kind of population growth pattern is characteristic of insects with short life spans, such as thrips, and most annual plants. The population grows very rapidly during a period of favorable weather (spring and summer), and then crashes rapidly when the weather changes in the autumn. Over several years, seasonal changes in a population of adult thrips living on roses show regular population peaks in the summer, with the crashes each fall occurring when cold weather comes (Fig. 5-3).

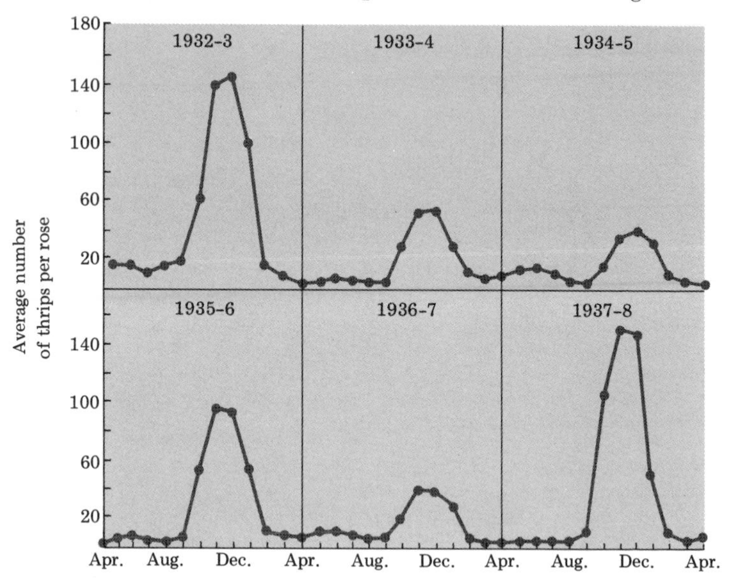

Fig. 5-3. Seasonal changes in a population of adult thrips living on roses.

S-shaped or Sigmoid Growth Curve

A more frequently encountered pattern of population growth is the S-shaped or sigmoid growth form, where growth starts slowly, accelerates rapidly in exponential form, and then decelerates and continues thereafter at a more or less constant level (Fig. 5-4). This growth curve is characteristic of larger organisms with longer life cycles and lower biotic potentials. Here, population growth is slow at first, in the **establishment** or **positive acceleration** phase. Then population increase becomes rapid, as in populations exhibiting J-shaped or exponential growth curves; this is the **logarithmic** phase. Somewhere in this period, an **inflection point** in the curve is reached where the population growth rate no longer continues to accelerate, but begins to decelerate. This **deceleration** phase is a slowdown of population growth caused by the gradual increase of the environmental resistance present in the system. The deceleration continues until a more or less equilibrium level is reached and maintained. This upper level, beyond which no major increase can occur, represents

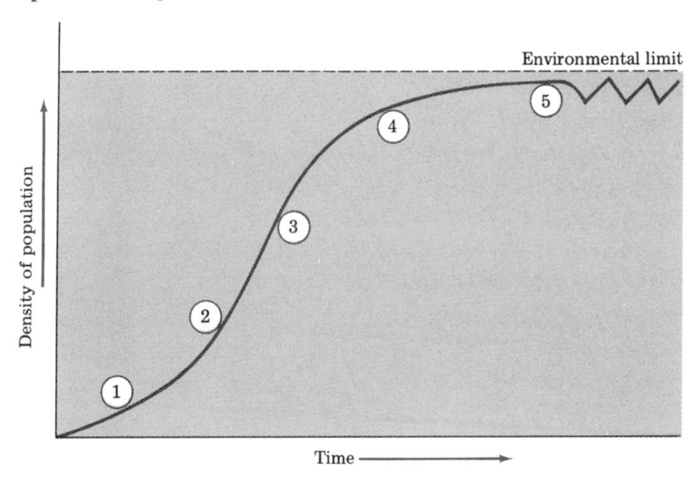

Fig. 5-4. The S-shaped or sigmoid growth form of a population. The indicated points along the curve are (1) the establishment of positive acceleration phase; (2) the logarithmic increase phase; (3) the inflection point, where the growth rate begins to slow down; (4) the deceleration phase; (5) the maximum population size or carrying capacity.

the size at which the population limits itself. Such a limiting size is often called the **carrying capacity** of the environment.

How is this carrying capacity determined? What factors control the maximum size to which a population can grow in a particular ecological setting? To consider these questions we must look at the ecological components of population regulation.

Population Regulation

As we have noted in discussing the two major types of growth form, limiting factors of some sort, which we have loosely called "environmental resistance," always slow or stop population growth. These factors can be of many kinds, but they are commonly grouped into two categories: **density-independent** and **density-dependent factors.** Any factor, whether limiting (negative) or favorable (positive) to a population, is density-independent (in terms of the total population) if its effect is constant regardless of the number of individuals in the population. It is said to be a density-dependent factor if the effect on the population varies with the density. Let us consider some examples of these components of environmental resistance.

Density-independent Factors

These factors exert the same effect on the population regardless of the number of individuals present. Climatic factors such as seasonal changes in photoperiod (daily amount of sunlight), rainfall, and temperature are almost always density-independent in their effect. Generally, such factors are most important in governing the population growth of small organisms (such as insects or plankton) with short life cycles and high biotic potentials. Weather and other physical environmental factors such as soil or water temperature are quite important in determining the length of favorable periods of annual growth for such organisms. They generally lack stability in their population size, which changes radically through the seasons.

The yearly abundance of the small sap-sucking insect *Thrips imaginis* in Australia is frequently cited as a classic example of density-independent population regulation. The biologists H. G. Andrewartha and L. C. Birch and their coworkers found that the thrips populations on flowers reach a minor peak of abundance in the southern winter (August), increase dramatically as late spring approaches (September to December), and then become very scarce until the next August. This is because in southern Australia flowers are abundant in spring but not during the long dry summer or in the growing period of the cool winter. With so many flowers, the abundant food and living space provide an unlimited environment for expansion of the thrips populations. They never have a chance to become excessively abundant on individual flowers because the long dry summer comes first and the thrips must leave the dying blooms, usually dying themselves in the search for new flowers. So there is not an absolute shortage of food, only a seasonal change in the accessibility of food. Hence the rather regular seasonal changes in the abundance of thrips each year (Fig. 5-3) depend upon the influence of the climate and upon the level of the thrip population at the start of the principal growing season.

Other density-independent effects may be caused by heavy rains, drought, human destruction of habitat, or even pesticide use. Nesting and hatching success as well as survival of young quail, ring-necked pheasants, rabbits, and other birds and mammals may be greatly affected by the timing of heavy rains, cold snaps, or even by lack of moisture. Muskrats and beavers are forced to migrate overland when drought dries up marshes or ponds and their burrow and lodge entrances are exposed. Flooding of their home environments, of course, can result in the same disruption of normal activities. During such overland movements, foxes, coyotes, and minks can easily attack the muskrats and young beavers. When man destroys extensive areas of natural habitat, as through clear-cut logging, draining of prairie potholes, and cultivation practices, the result is serious reduction or extinction of native animal populations regardless of their density. The widespread use of pesticides has caused dramatic declines in wildlife populations independent of the number of individuals present at the time of spraying. Robin populations in sprayed areas of Wisconsin dropped 69 to 98 percent below those in unsprayed areas. Their earthworm diet was responsible; the earthworms had built up a concentration of DDT from feeding on leaves that decayed in the soil beneath elm trees sprayed to control Dutch elm disease. The degree of mortality caused by the poison was independent of the size of the robin populations.

Density-dependent Factors

These factors exert effects varying proportionately to the size of the population. Density-dependent factors are responsible for the steady-state population sizes seen in the upper-level portion of the sigmoid growth curve. These kinds of factors operate throughout population growth, not solely at one point in time. They normally begin to act well below the carrying capacity (causing the inflection point in the sigmoid growth curve) and intensify their effect as the upper population limit is approached. There, at the equilibrium density, the production of offspring exactly balances the loss of adults by death or emigration. Should the population size exceed the equilibrium density, the density-dependent factors exert a stronger effect and cause a greater rate of loss. This brake operates until the equilibrium density is reached again. Likewise, if the population size drops below the equilibrium density, the density-dependent factors will permit the population to increase. Since growth and reproduction require time, it is obvious that there will be a time lag between cause and effect, and hence the population size will oscillate about the equilibrium density. As a governor mechanism regulates the speed of an engine, so density-dependent factors regulate population growth, preventing overpopulation in most animal and plant species. Thus they, rather than density-independent factors, are said to be the factors primarily responsible for the achievement of a steady-state or mildly fluctuating density level at the carrying capacity of the environment.

Density-dependent factors are usually biotic in character, involving interactions with other organisms. **Intraspecific competition** is commonly an important regulating process in natural populations. Individuals of the same species compete for a resource in limited supply; under such conditions the population can accommodate only a certain maximum density. Food and space are normally the environmental resources in short supply. When the rate of reproduction is excessive and the equilibrium density is surpassed by new animals added to the population, competition for food or space will cause an increasing mortality rate, which accelerates as the population increases. The younger animals are usually most affected and hence the addition of new reproductive animals to the population is curtailed, eventually dropping the birth rate back to a level that yields an equilibrium density.

Density-dependent agents operating on a population may also include parasites, **pathogens** (bacteria, viruses, and protozoans) causing disease, and predators. Parasites and pathogens spread more easily from organism to organism in high-density populations of either plants or animals and hence tend to reduce the density of the hosts through mortality. Predators can find prey more easily in populations of high density and as long as they are not swamped by extremely high numbers of potential prey, they operate as a very efficient density-dependent factor.

Emigration also acts as a density-dependent process. At high population levels, emigration of a portion of the adults or young into unoccupied territory (when available!) can regulate the population density of the original area. Each year in black-tailed prairie dog towns of the Rocky Mountains, the adults of a family group will leave their home territory in a town and emigrate to the outskirts of the town to colonize a new area. In this species, the younger, inexperienced animals are able to remain behind in the center of the town in familiar territory while the mature, experienced prairie dogs take on the more dangerous task of establishing residence on the edges of the town. Over the years the population density in the town's center remains more or less constant.

Other processes operating in a density-dependent fashion include physiological and psychological control mechanisms. The social stresses caused by crowding have been shown to act on the individual through the endocrine system. As local populations of certain locust species increase in numbers, at a certain density crowding causes a hormonal change in the new young locusts, so that they develop longer wings and other morphological changes that will equip them for migratory movement. When this new generation matures, the whole series of local populations in an area will move off in flight *en masse,* migrating hundreds of miles in huge swarms. In vertebrates, these endocrine-induced changes are particularly associated with the pituitary and adrenal glands. Increasing population density leads to inhibition of sexual maturation, lowered sexual activity, and inadequate milk production in nursing females. Stress caused by crowding also curtails the number of offspring produced per litter, through spontaneous abortion or absorption of tiny fetuses into the uterine walls in pregnant mammalian females.

In general, species controlled by density-dependent factors tend to be larger organisms such as birds or mammals or trees. They tend to have longer life cycles, usually more than a single year. Their rate of production of offspring is lower and thus they have lower biotic potentials and more stable population sizes than populations controlled principally by density-independent factors.

Of course, these various regulating factors we have been examining are *not* perfect, and no animal or plant population maintains a perfectly constant density over time. Population fluctuations are a basic fact of life. Before closing our consideration of population growth and regulation we should look briefly at the several classes of fluctuation in density.

Population Fluctuations

Periodicity in populations is commonly seen in **seasonal** or **annual** changes in density. These are often correlated with corresponding seasonal or annual variations in physical limiting factors such as temperature and rainfall.

In the temperate zones of the Northern and Southern Hemispheres, populations primarily respond to seasonal changes in temperature. One of the most noticeable responses in many species is **hibernation,** in which an animal greatly slows its metabolic processes and remains quiescent during the winter months. Frequently, the period of hibernation is spent in an underground burrow or den where the inactive animal is protected from disturbance and where a relatively constant environmental temperature may be maintained by trapping the escaping body heat. During the summers in desert or seasonally hot areas, some animals escape the heat by **estivation,** a state of reduced metabolism and activity similar to hibernation but carried out to survive a hot period rather than a cold period. Leafing of perennial trees and shrubs in the spring and leaf-drop in the fall are other seasonal responses to temperature changes in temperate climates.

In the tropical areas of the world, rainfall is commonly the key factor governing seasonal changes in population structure and functioning. Many organisms pass the tropical dry season in state of **diapause** or dormancy, hidden underground away from the desiccating temperatures and low humidities characteristic of this time of year. The first rains of the wet season give the cue for breaking of diapause and these organisms become active again. Annual plants spend the dry season in the form of dormant seeds. Leaf production and leaf-drop in tropical deciduous forests are tightly correlated with the alternation of wet season and dry season. Termite and ant swarming of reproductive adults (winged queens and males) from nests is often cued by rainfall.

Obviously, then, in both tropical and temperate environments population growth is keyed to favorable periods in the annual cycle. During periods of environmental stress in the course of a year the visible population density may drop drastically as the animals seek shelter for a dormant stage or a portion of the population dies.

Long-term oscillations in density, however, are not obviously related to seasonal or annual change and may involve remarkably regular cycles of abundance with many years between peaks and depressions. The best known cases are described from arctic mammals, but periodicity of this nature exists in certain birds, insects, and fish, and seed production in certain plants.

A classic case of a ten-year oscillation is that of the snowshoe hare and the Canadian lynx, a large relative of the more familiar bobcat. Since about 1800 the Hudson Bay Company in Canada has kept records of the number of pelts of fur-bearing mammals taken by trappers and purchased by the company each year. When these data are plotted over a long span of years (Fig. 5-5), it is apparent that the lynx reaches a population peak every nine to ten years (average of 9.6 years). These peaks of abundance are often followed by "crashes," or rapid declines in number, and the lynx becomes very scarce for several years. When numbers of pelts of the showshoe hare are plotted for the same time period, it is found that the hare follows the same general cycle, with a peak abundance usually preceding that of the lynx by a year or more. Since the lynx is largely dependent upon the hare for food in these arctic areas, it is obvious that the cycle of the predator is related to that of the prey. But the two cycles have not as yet been shown experimentally to be related by cause and effect.

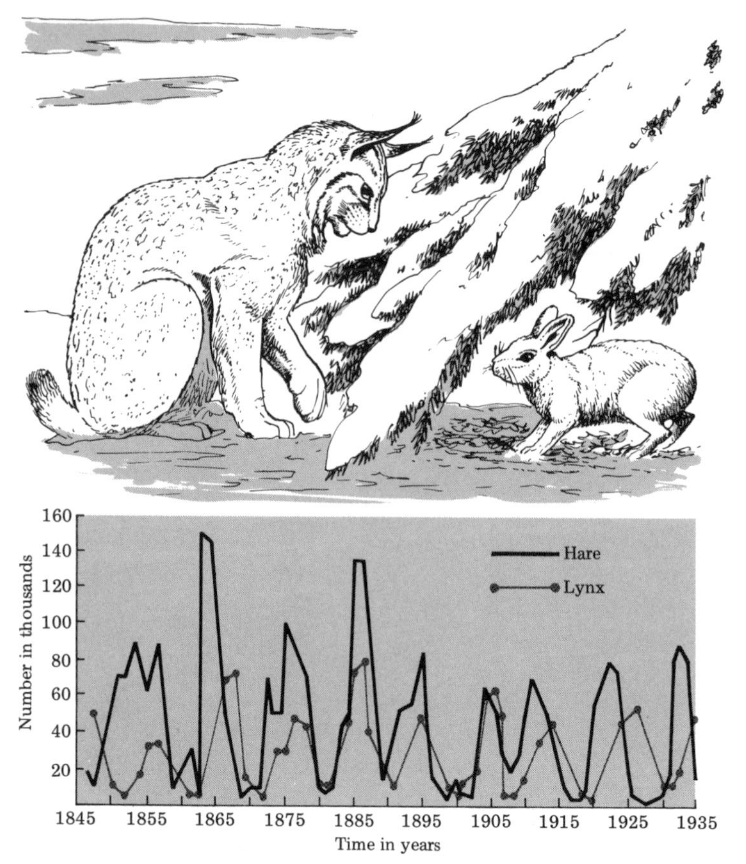

Fig. 5-5. Changes in the abundance of the Canadian lynx and the snowshoe hare, as indicated by the number of pelts received by the Hudson Bay Company from 1845 to 1935. This is the classic case of long-term cyclic oscillation in population density.

A shorter, three- or four-year cycle of abundance is characteristic of many northern mice, lemmings, voles, and their predators such as the snowy owl and the arctic fox (Fig. 5-6). Lemmings live in the northern tundra of Europe (two species) and North America (two species), and become very abundant every three or four years, then crash to very low population levels within a single season. The populations of the predators likewise increase and then crash, following the sudden decline of the lemming populations. The arctic foxes compete for the remaining food following a lemming population crash and hence may starve, but the snowy owls respond to the decline in lemming numbers by migrating far to the south in search of food, occasionally even to North Carolina. Bird watchers in the United States can count on this migration occurring every three to four years. The regulating factors causing the lemming population crashes appear to be density-dependent in nature, involving pathogens, changes in

Fig. 5-6. The brown lemming (*Lemmus trimucronatus*) and its two principal predators, the snowy owl and the arctic fox. The graph shows the short-term, three-to-four-year peaks and lows of the lemming cycle on Victoria Island (eastern end), Canada.

food quality and quantity, and physiological and psychological stress as well as predators. After such a crash, the lack of predators and the low density of lemmings in relation to the amount of available food and space allow the lemming populations to increase almost unchecked again for the next several years. Then the overcrowding again produces a sudden population crash and another complete cycle has occurred. Occasionally, dramatic emigrations occur to relieve exceptional outbreaks, as has been mentioned earlier in discussing population dispersal (Chapter 4).

In some insect groups, long-term population cycles appear to be utilized as a successful strategy to avoid predation. The seventeen-year cicada has a seventeen-year gap between adult generations. In the year of an emergence, the adults emerge synchronously in early summer and live only four to six weeks. These insects come from nymphs that hatched originally from eggs laid over a period of several weeks some seventeen years earlier. Yet the developing nymphal stages in the soil manage to keep track of the years and synchronize their emergence from the ground to become adults at exactly the same moment as the rest of the population. Until the cuticle (outer skin or integument) becomes hardened by exposure to the air for several hours, the newly emerged adult cicada is helpless (Fig. 5-7). However, because the population emergence is synchronized the predators in the area cannot destroy the millions of individuals that suddenly appear. Thus

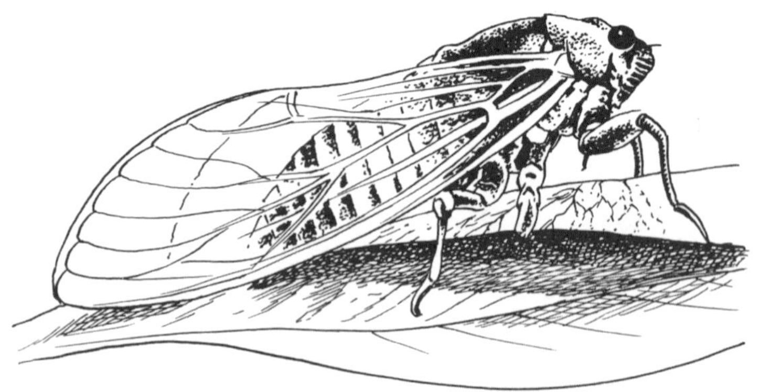

Fig. 5-7. A periodical cicada, an insect with a seventeen-year life cycle.

the seventeen-year cicadas overwhelm predators when they appear. The extreme length of the cycle makes it very difficult for a predator to get into a similar cycle and specialize in cicadas. What would the bird or wasp eat the other sixteen years?

It is also hypothesized that the long cycle enables the cicada to escape certain fungal pathogens present in the soil of Eastern deciduous forests. The spores of these molds may be unable to survive until the next emergence of the adults. Naturally, the other density-dependent factors operate during the years the cicada nymphs are underground. Intraspecific competition for nymphal food (the cicada nymphs feed on sap in tree roots) and annual predation on nymphs by moles and shrews probably play an important role in establishing the carrying capacity for this cicada in the areas it inhabits.

Summary

Population density refers to the number of individuals per unit of space and may be measured in several ways (crude density, relative or ecological density, and index of relative abundance). Growth of overall number in a population involves natality, mortality, and survivorship. The form of survivorship curves depends on the time in life when mortality factors act. Should all environmental restraints be absent, all individuals would survive to reproductive age at each generation and the population of that species would have reached its biotic potential (simply maximum natality less minimum mortality). The fact that natural populations do not increase at this maximum possible growth rate reflects the strength of the environmental resistance present. The result of the interaction of the biotic potential and environmental resistance is a characteristic pattern of increase in numbers called the population growth form. The two extremes of this form are the J-shaped or exponential growth curve and the S-shaped or sigmoid growth curve; the latter levels off at the carrying capacity of the environment. Population regulation is accomplished by a variety of density-independent and density-dependent factors. Commonly, one will find periodic fluctuations in population density relating to seasonal or annual variation in limiting factors of the physical environment. Relatively long-term oscillations that cannot be explained as easily are found in arctic mammals and other species.

6
THE ECOLOGY OF COMMUNITIES

Individuals constitute populations, and populations in turn make up plant and animal **communities.** Thus a biotic community is an assemblage of populations of various species living in a particular region or habitat. It is the *living* part of the ecosystem. Communities have internal structure and regulatory processes just like the population and the individual organism. We have already touched upon many of these basic structural and energetic interrelationships in earlier chapters, without making special note that they are most significant on the level of the community. It is relatively meaningless to speak of energy and element flow, food webs, interspecific relationships, and so forth outside the context of the total biotic community. Thus in this chapter we will look specifically at several additional concepts of community ecology that will be of special importance in relating basic ecological phenomena to population growth in man and the resultant crises in environmental deterioration.

Differentiation in a Community

The word *community* can describe biotic assemblages that differ greatly in size, from the living organisms in a decaying log to the fauna and flora of a huge rain forest. **Major com-**

munities are those that, together with their physical habitat, form more or less complete and self-sustaining ecosystems. In such systems solar energy is the principal or only needed input from external areas. Such communities are relatively independent of adjoining communities because of their sufficient size and level of organization. **Minor communities,** on the other hand, are more or less dependent on neighboring aggregations of organisms for an input of energy. The decaying log or a temporary rain pool in a desert canyon represent minor communities with comparatively short but intense periods of biological activities.

The various plant and animal species that compose a community of any size have different ways of life. Hence community differentiation shows simultaneously on the horizontal plane, on the vertical levels, and temporally. **Horizontal differentiation** is the patchy or mosaic-like occurrence of species; we have already seen it in discussing the dispersion of populations (Fig. 1-3). **Vertical differentiation** shows in the different heights to which plants of various growth forms ascend and in the resultant "layering" of animal species living at these several levels in the community (Fig. 1-2). **Temporal differentiation** is exhibited by different species carrying on similar functions at different times in both daily and seasonal cycles (Fig. 1-4). Nocturnal owls and diurnal hawks both hunt small rodents, but their periods of activity do not overlap. These aspects of community differentiation are not unique to terrestrial communities. We may see the effects of environmental factors on distribution of aquatic life forms particularly well in ponds and lakes.

Lake Communities

Lakes are large, relatively still bodies of fresh water. They may be formed by glacial erosion, uplifting of mountains, deposition of debris in the beds of slow-flowing streams, or other geological activity. With dams on large rivers, man has created many new lakes for power, water storage, and agricultural irrigation.

Within any lake there are notable gradations of light penetration, temperature level, and oxygen content of the water. The depth to which light will penetrate is limited by suspended materials, such as silt particles; hence the extent of turbidity restricts the photosynthetic zone in the aquatic habitat. For this reason phytoplankton tend to occur in the upper levels of the lake. The annual cycle of seasonal changes in water temperature also affects the distribution of plant and animal life within the lake. As the surface layer warms up from more intense sunlight in the spring and summer, the difference in density between the surface water and the deeper, colder layers establishes a barrier against mixing. The nutrients in the sediment on the lake bottom are no longer lifted by upwelling of the deep water, and the more highly oxygenated surface waters do not sink. The overall result of this pronounced thermal stratification is to halt mixing of the nutrients, the dissolved oxygen, and the plankton near the surface with the deeper water. Hence most fish have to move upward in the lake during the midsummer period to gain sufficient oxygen and plankton. In the fall this thermal stratification disappears again and free mixing of the shallow and deep water can take place.

With erosion from surrounding slopes and input of sediment by rivers, smaller lakes tend to support relatively short-lived communities of aquatic life. The natural succession that takes place as the lake basin fills in eventually leads to a pond, a marsh, and finally a grassy depression in the landscape. Man's giant reservoirs in the American southwest, as well as in other parts of the world (for example, the Aswan Dam of the United Arab Republic), have an especially short lifetime because the particular rivers dammed to create the reservoirs carry large loads of silt.

Dominance in a Community

When one looks at a community of organisms from an ecological viewpoint, one or several species usually stand out as major influences on the composition and internal energy dynamics of the community. Such species are said to be **dominant** because of the *numbers* of organisms present, their *size*, or their *energetic* activities. Thus out of the hundreds of species in a California pine forest, we immediately notice the greater numbers of Ponderosa pines and regard that species, *Pinus ponderosa*, as the dominant organism in the community (Fig. 6-1). In a fifty-acre pasture in the eastern United States, we might find the following numbers of animals and plants present in the total area covered:

Fig. 6-1. A Ponderosa pine forest community in the western United States, showing *Pinus ponderosa* as the readily identifiable dominant species.

dairy cattle	50 individuals
turkeys	6 individuals
sheep	1 individual
bluegrass	45 acres
white clover	3 acres
oak trees	2 acres

Clearly, the dairy cow is the dominant species among the consumers and the bluegrass is dominant among the producers in this farmland community.

Productivity and Associated Concepts

We frequently want to know how much an area in a particular type of biotic community will produce. For instance, agriculturists use the concept of **productivity** to describe how many cattle can be raised per hundred acres of pasture. Wildlife biologists can likewise use it to describe the number of native antelope subsisting on a hundred acres of savanna in Kenya. Applied ecologists may then desire to compare the production of such man-maintained communities and natural communities. The concept of productivity provides a way to standardize such comparisons. Productivity is defined as the production of organic matter per unit of area over a particular time interval, usually a year. Productivities measured on an annual basis take into account the differences in organic matter present at various times of the annual cycle, and hence prevent the erroneous conclusions that may be reached from comparisons of two regions sampled at different seasons.

Sometimes, however, it is useful to know the number of individuals or the total biomass present in a community at one particular time. Such a measurement is termed the **standing crop** of that area. During the winter in temperate-zone localities, the standing crop would be small. But in the late spring and summer, as growth and reproduction take place, there is a very considerable increase in the amount of organic matter making up the biomass of the community. Similarly, in the tropics the forests have a lower biomass in the dry season than in the wet season when the trees produce more foliage in response to heavier rainfall.

Finally, ecologists speak of the carrying capacity of a community or area, a concept we have already mentioned in Chapter 5. This is the largest standing crop of a species that an area can support without deterioration of the habitat. Should the carrying capacity be exceeded through overpopulation, the various regulatory factors we have previously discussed will soon bring the standing crop below the critical level again.

Succession

Change in a biotic community is more or less constant. New species enter the area; old ones find conditions changing and must move or die out. The continual replacement of one community by another is called **succession,** and it terminates only when a **climax** stage is reached. The process of succession is partly a result of ever-changing climate and physical habitat in the region, and partly of the activities of the previous community's organisms. Lichens, for instance, are an excellent **pioneer** organism in the first organic community growing on a bare rock surface. Their secretion of organic acids breaks down the rock surface, making available enough nutrients and substrate * for moss spores to germinate and establish a new rock community dominated by mosses. In turn, mosses trap more rainwater than do lichens and in general change the environmental conditions so much that many of the lichens die out. Thus a new stage in succession has been attained. Next, the rich bed of mosses may provide satisfactory conditions for the establishment of ferns. This series of steps or communities that develop in the process of succession is called a *sere* (Fig. 6-2).

After passing through relatively transitory communities (called **seral stages**), the sere ends in a **climax community**

* The substance or base on which an organism grows.

Fig. 6-2. Forest succession from annual weeds to mature forest. Note the increasing stratification as one seral stage advances to another. (*For credit see p. viii.*)

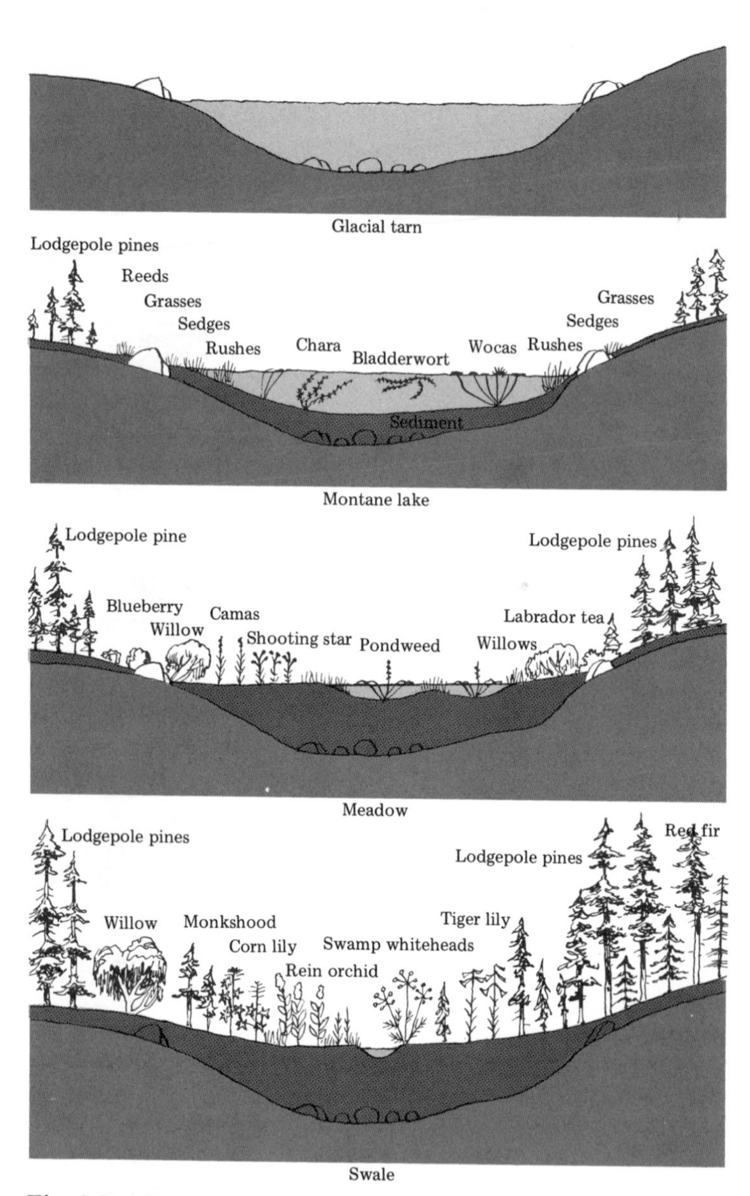

Fig. 6-3. Stages in primary succession from a lake to climax forest in the Sierra Nevada mountains of California.

that is best suited for the prevailing climate or soil conditions. If the succession that led to this climax stage began on an area such as newly exposed rock or sand not previously occupied by a biotic community, we have **primary succession** (Fig. 6-3). The first seral stage in primary succession is called a pioneer community. Should community succession proceed in an area where a community was removed (e.g., a plowed field or clear-cut forest), we have **secondary succession.** The process of secondary succession is usually quite rapid, as nutrients, soil bacteria, and other necessities are already present from the activities of earlier organisms. Thus weeds spring up in great numbers in a plowed field within a few weeks; they need not wait for the early stages of primary succession to prepare the soil substrate for the specific requirements of these evolutionarily higher flowering plants. Shrubs and, later, trees will follow the weed seral stage if the climate and soil are suitable. The whole trend in either primary or secondary succession is toward a stage with a complex community food web where less energy is wasted and there is high ecological efficiency at all levels of the food pyramid.

Major Climax Formations

The major climax community types of the world are called **biomes.** Biomes are general groupings of similar communities. These groupings are based upon the distinctive life forms of the important species in each community. Thus the great number of biotic communities around the world is reduced to a concise and mentally manageable set of nine or so biomes.

In terrestrial regions, dominant vegetational types are usually used to characterize biomes. The principal biomes of the world are:

Tundra
Coniferous forest or Taiga
Temperate deciduous forest
Chaparral
Grassland
Desert
Tropical savanna
Tropical deciduous forest
Tropical rain forest

Tundra

North of the Arctic Circle extends a vast treeless region with a very short growing season (June–July), extremely low temperatures during the winter and cold temperatures even during the brief summer, and a permanently frozen soil (the **permafrost**) below the surface. This is the arctic **tundra** biome (Fig. 6-4). Under such harsh conditions only a few species of animals are found, but their populations often reach huge numbers. Human visitors to tundra areas often find mosquitoes to be overwhelmingly abundant organisms in the arctic summer.

Most of the mammals are burrowing species, such as lemmings and mice. Some of the larger mammals, such as caribou and musk ox, must migrate annually in search of food. These herbivores feed in the tundra zone only during the short time of production of grasses and lichens. Other migrants, especially the birds, depend for food upon smaller organisms that are only active during the warmer summer months.

Fig. 6-4. The tundra biome found north of the Arctic Circle and on mountain tops above timberline, here seen near St. Mary's Glacier in northern Colorado.

Plant life is represented by many lichen species, mosses, and low-growing flowering plants such as sedges, grasses, and hardy shrubs, growing in a tight mat upon the surface of the ground. Food chains and webs are relatively simple in the Arctic and for that reason they have been well studied by ecologists. There is a great amount of annual moisture in the tundra, but much of it is tied up in snow or permafrost or lost by evaporation. Thus relatively little water is available to tundra organisms.

Coniferous Forest or Taiga

South of the tundra, around the polar region of the Northern Hemisphere, one finds a broad band of evergreen coniferous forest, or **Taiga** (Fig. 6-5). In the Southern Hemisphere only a small area of coniferous forest exists. This results from the lack of land area, since the proper latitudes largely fall between the coast of Antarctica and the tips of the southern continents. The spruces, firs, pines, and other conifers of the Taiga are able to grow at these latitudes because of the increase in *available* annual moisture. This moisture increase in the Taiga as compared to the tundra results from warmer temperatures farther away from the poles. In these higher temperatures, large amounts of moisture cannot remain in frozen forms that cannot be absorbed by plant life, as in the tundra. Although the conifers are subject to long cold winters at these latitudes, they remain green all year and do not drop their leaves (needles) simultaneously. Yet the stresses of the subarctic winter bring photosynthetic production to a virtual halt. Flowering plants like willows and birches grow along streams and in pioneer habitats in the Taiga, and drop their leaves in the winter. Animal life in this biome is also relatively poor in species. Fur-bearing mammals such as bears, weasels, and martens are typical inhabitants. Predators are not as likely to be migratory in this biome as in the tundra since the prey species that inhabit the Taiga can be active above ground during the winter.

Temperate Deciduous Forest

Rainfall in this biome is sufficient to support broad-

Fig. 6-5. The coniferous forest (Taiga) biome in Colorado.

leafed shrubs and trees. The great hardwood forests of eastern North America (Fig. 6-6) and Europe are excellent examples of **temperate deciduous forest.** Since this forest type occurs in the temperate zone, the species are subject to considerable temperature fluctuations and a cyclic alternation of photosynthetic production, which is high in the summer months and low in winter. This biome was the cradle of European civilization, and historically most scientists, including those living today, have worked in it. For this reason, much of our ecological thinking is biased in the sense that it is based upon experience in the temperate zone rather than in the tropics. In some cases this has probably been quite beneficial. Linnaeus probably succeeded in establishing a binomial system of nomenclature for plant and animal species in the eighteenth century because of the *few* organisms he had around him in Sweden where he lived. Had he resided in tropical South America, the incredible abundance of different kinds of organisms might have overwhelmed him and prevented such a logical and adaptable taxonomic system from coming to mind.

Fig. 6-6. The temperate deciduous forest biome in western New York.

Chaparral

The Mediterranean climate of parts of southern Europe, southern Australia, and southern California is very dry for most of the year, with low total precipitation in the form of winter rains and occasional summer thundershowers. The low, shrubby vegetation in these areas has thick, evergreen leaves whose toughness and waxy coatings make them resistant to prolonged periods of drought. This **chaparral** biome vegetation is also **fire-adapted**; that is, it depends upon periodic fast-burning fires to perpetuate itself and allow reproduction of the various species of shrubs at the expense of trees. After a fire passes, with the first rains of the next growing season the chaparral species sprout from their stumps or from seeds lying dormant in the soil. However, if fires do not race through the chaparral every ten to twenty years, the community becomes overly mature and growth of new seedlings is prevented by the densely growing older plants. When a fire finally occurs, it will be extremely destructive because of all the accumulated fuel under the older plants (old leaves, branches, etc.). Even the stumps and roots of many shrub species will be killed by the intense heat. When shrubs have been destroyed this way, tree seeds floating in from adjacent forest areas can germinate and grow in the area before the chaparral can recover. Thus chaparral is an example of a fire-maintained climax community.

Grassland

In the extensive temperate plains areas of North America, Eurasia, and Australia there is sufficient rainfall to support a heavy growth of herbaceous plants, especially bunch grasses and spreading grasses (species with underground stems). No trees are present except along watercourses. Burrowing rodents are common and often form extensive colonies. Large herds of grazing animals (including marsupials such as kangaroos in Australia) were typically found in the grasslands of recent times; today, many of these species have disappeared or been nearly exterminated by man. Man has also been interrupting the ecology of grasslands by cutting through the sod with plows and other mechanical devices. Today, little remains of the flora and fauna of the original Great Plains and prairie areas of the eastern United States. In fact, many of the grasses would probably have disappeared under the farmer's plow long ago were it not for their accidental preservation along railroad track right-of-ways. The great herds of American bison were reduced by hunting to a few individuals in Canada by the end of the last century, but fortunately this remnant was bred to reintroduce the species to United States park areas.

Fig. 6-7. The desert biome, here represented by the Mojave Desert in eastern California.

Desert

The **desert** biome (Fig. 6-7) is worldwide in extent and is characterized by very low rainfall, a high evaporation rate, high diurnal temperatures, and low nocturnal temperatures. The plant species present are generally quite low in profile. Desert shrubs and other plants tend to have reduced, waxy, drought-resistant leaves. Some plants such as cacti have virtually lost their leaves and depend upon greatly expanded photosynthetic stems for **autotrophic** * food production. Primary production in a desert is low and hence primary consumption by animal life is low. Relatively few species of animals are active during the daytime or even live in the desert. Because of the cool nights, dew is a significant source of moisture for desert organisms, which lack the availability of surface water or frequent rain showers. Many desert plants

* Capable of producing its own food. An autotroph, like a typical green plant, does not have to depend on other organisms for food. It acquires carbon dioxide from the air and water from the soil, and from these inorganic molecules it makes, through photosynthesis, the sugar glucose, which can be used in its cells as food.

can absorb dew and translocate its moisture from their leaves to their stems and roots.

Tropical Savanna

In some of the tropical areas of the world, one finds grassland with scattered trees. This mixture of dry-adapted herbaceous plants, shrubs, and occasional trees is called **tropical savanna.** In overall appearance (Fig. 6-8) it is like an open park. Acacias in Africa and palms in South America are the usual trees on the savanna. Rather large mixed herds of

Fig. 6-8. Two views of the African savanna.

primary consumers characterize the savanna fauna. In eastern Africa, for instance, gazelles graze the lowest layers of the vegetation while giraffes feed on the tops of the thorny acacias. Large predators such as lions prey primarily upon such herds of herbivores. In tropical Australian savannas the primary consumer herds are marsupials.

Tropical Deciduous Forest

Where the rainfall becomes heavier on the margins of a tropical savanna and can support regular forest, one finds the **tropical deciduous forest** biome (Fig. 6-9). This biome is subjected annually to a dry season–wet season environmental cycle instead of the cold winter–hot summer regime of its temperate-zone counterpart. With uniformly warm temperatures all year, the growth of the trees and herbaceous plants of this tropical forest is affected only by seasonal distribution of the rainfall. Hence, in the annual dry season such forests lose most leaves and then leaf out again at the onset of the rainy season. In the considerable number of species in such a forest, most exhibit special adaptations for passing the dry season in much the same manner as temperate-zone species pass the winter.

Tropical Rain Forest

In the circumtropical biome of the **tropical rain forest,** rainfall tends to be more or less evenly distributed throughout the year. The mean annual temperature lies between about 68°F and 82°F, with very little seasonal fluctuation. In fact, the change from daytime to nighttime temperatures in a rain forest is greater than the annual mean temperature fluctuation. Neither temperature nor water is a limiting factor on plant or animal growth. Soil conditions do exercise some limitation on plant growth. Biotic interactions are very important as limiting factors; there are more different kinds of organisms in this region than in any other biome. Such unusual creatures as parrots, army ants, and tapirs (Figs. 6-10 and 6-11) are representative of the diversity of life in the tropical rain forest. Yet there are fewer individuals of any one species per unit area than in other biomes.

Fig. 6-9. The tropical deciduous forest biome during the dry season in northwestern Costa Rica.

Fig. 6-10. The tropical rain forest on the Osa Peninsula in Costa Rica, the richest and most diverse climax community on earth. Parrots and large army ants are typical inhabitants.

Fig. 6-11. A South American tapir, a large mammal found in tropical rain forests.

Fig. 6-12. Adaptive features of plants in a rain forest. (A) A bromeliad with a rosette of basal leaves forming a basin for rain water. (B) A "drip tip," the structurally modified end of a typical tree or shrub leaf in the rain forest.

Arboreal epiphytic plant life is particularly well developed in Central and South American rain forests. Orchids and "tank plants" of the family Bromeliaceae are exceptionally abundant. The latter plants catch rain water in the leafy basins formed by their rosettes of leaves (Fig. 6-12), and aquatic communities of insects, frogs, and other organisms develop in these pools. The rain forest canopy becomes essentially continuous in some places so that the forest floor receives little light; hence undergrowth is limited. Where the canopy is broken, the forest is much more complex and trees of various heights present a layered or storied picture. Long hanging vines called lianas cascade down from the canopy in such areas. Most plants are evergreen and may exhibit "drip tips" on the ends of their leaves, apparently an adaptive feature to speed runoff of rain water (Fig. 6-12). Because of the constant high humidity, a miniature plant community composed of mosses and fungi can develop on the surface of a broad, wet leaf of a tree or shrub within a short time. These tiny plants making their home on the surface of leaves are called **epiphylls.** Such epiphylls would cut down on the photosynthetic capability of that leaf (and hence of the shrub or tree) in the dim lighting of the understory of the forest. The drip tips of many plant species are a distinct advantage in protecting their leaves from the development of an overgrowth of epiphylls.

Climax Communities: Their Importance

Terrestrial climax communities tend to be more stable and have more diverse paths of energy flow in food webs than do intermediate seral communities. Hence they are generally the most productive of the natural communities in a sere. They are composed of the plants and animals best suited by evolutionary adaptation to the prevailing climatic and soil conditions of the region. Today, climax communities everywhere are being threatened by human activities. If a climax community is destroyed, we know that hundreds of years are necessary for succession to create another. Thus it becomes imperative for us to be fully cognizant of the risks involved in any major environmental alteration. We shall be looking at the results of some of these alterations in the next chapters.

Summary

A biotic community is an assemblage of populations of various species living in a particular area. Within this mixture of organisms one finds horizontal, vertical, and temporal differentiation. Dominant species stand out because of their numerical abundance, size, or energetic activities. The productivity of a community is measured by the production of organic matter per unit of area over a specified time interval, usually a year. The total biomass or number of individuals present at a particular point in time is the standing crop of a community. The carrying capacity of an area or community is the largest standing crop of a species that an area can support without ecological damage. The process of continual replacement of one community by another is called succession and occurs because the activities of the previous inhabitants change the environmental conditions. The series of steps or communities that develops during succession is called a sere. Primary succession begins with a pioneer community and ends in a climax community. Secondary succession proceeds in an area where a community was recently removed. The major climax community types of the world are called biomes and include tundra, Taiga, temperate deciduous forest, chaparral, grassland, desert, tropical savanna, tropical deciduous forest, and tropical rain forest. Climax communities are the most stable and most productive of the natural communities in a sere, and take hundreds of years to replace if destroyed by environmental alteration.

7 ENVIRONMENTAL ALTERATION

We have seen that communities are assemblages of plants and animals sharing a common environment and operating as a closely interrelated unit. Despite the diversity of life on earth, the number of major kinds of communities is small; and after viewing these principal biomes in Chapter 6 we realize that each is the climax type for a particular set of environmental factors, being principally dependent upon available precipitation and annual temperature cycles. In each case, the climax community, arising through a series of intermediate communities, is the result of the succession process. We noted that when the sere reaches the climax stage, things become relatively stable; that is, energy flows are more balanced, the mature community is more diverse, and there is less chance of catastrophic changes in the biotic balance, should one species suddenly boom in numbers or become extinct.

With this review as background, why is succession so significant when we consider factors that endanger natural communities? Simply because most of the natural communities that are particularly threatened are the *climax* communities in that area. *When a climax community is destroyed, the only way to restore it is by the lengthy process of succession.* It may take hundreds or thousands of years for a sere to reach that final stage; thus once a mature community is altered, the damage is almost permanent on the time scale of human activity.

With this fact in mind, let us view a selected spectrum of factors that change, pollute, or otherwise threaten natural communities. Underlying perhaps *all* of these problems is the basic factor of the burgeoning population and technology of man, which we shall treat later in detail.

Fire

Fire is a natural phenomenon that can happen in almost any terrestrial ecosystem because of the accumulation of plant remains or plants that at some point in the year become combustible. Under these circumstances, lightning or acts of man can set off uncontrolled fires. But though these fires are destructive in the immediate sense, there is no reason to think of fire as an evil to be totally abolished by man's efforts from all ecosystems. In the United States our thinking on the place of fire in the natural forest has changed radically over the past hundred years.

Historically, great fire disasters in northern lumber towns and forests in the nineteenth and early twentieth centuries made the American public consider fire in natural communities inherently undesirable. American foresters trained in European schools imported the protective concepts that were useful in the intensely tended forests there. Thus for decades the U.S. Forest Service emphasized in a well-advertised campaign that fire was extremely bad for the forests of America. However, ecologists have recently shown that where fire has been a natural part of the forest environment, it is frequently necessary for effective forest management.

In the southeastern United States, for example, fire is necessary for successful reproduction in longleaf pine forests. The longleaf pine grows along the coastal plain from Virginia south to Florida and west to Texas. In their natural state longleaf pines are well spaced and there is a light cover of grasses on the forest floor. When these forests were first logged, reproduction of the longleaf pine was very poor. At first, ground fires were blamed; so fire was tightly controlled or totally prevented in such forest areas. Soon there was no reproduction of the pines at all. Upon investigation of the

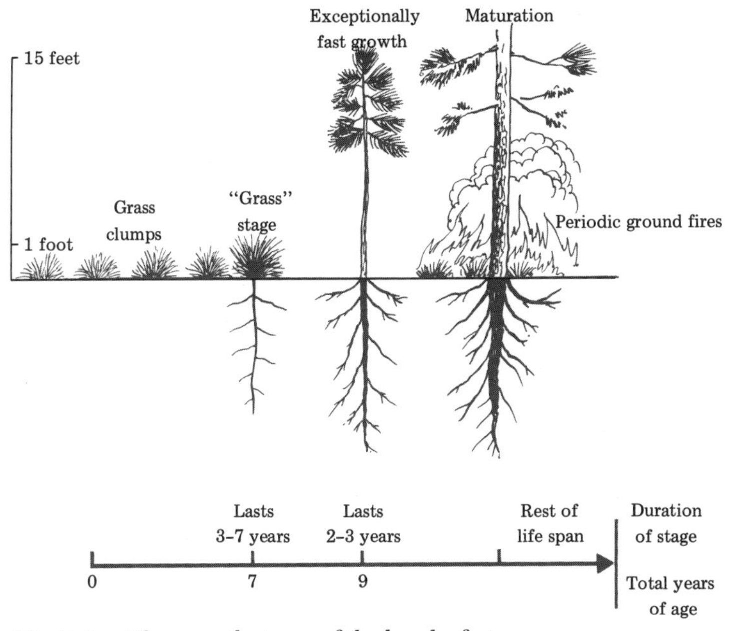

Fig. 7-1. The growth stages of the longleaf pine.

species' life history, the answer to this problem became evident (Fig. 7-1). For the first three to seven years after sprouting, the young longleaf pine consists of a long taproot, with a thick terminal bud, protected by a long tuft of needles, above ground. Grass fires sweep by too rapidly to harm such a bud. During the next two or three years, the pine grows exceptionally fast, to ten to fifteen feet in height, and thus raises its more sensitive new branch growth above the range of the normal grass fire. The development of a thick, fire-resistant bark protects the young tree from further ground fires. The periodic ground fires characteristic of southern forests are necessary to prevent the incursion of much less valuable hardwoods or other species of pine into the longleaf pine stands. Without the fires, the fire-susceptible hardwoods are able to grow relatively well and shade out the young longleaf pines. Thus fire is *necessary* to maintain the economically desirable climax community of longleaf pine. Today this is recognized by forest managers, and use of periodic controlled ground fires is a common practice.

Agricultural Clearing

Historically, the clearing of land for cultivation has been the major means of landscape alteration. But currently the United States, the Soviet Union, Ireland, and Japan, among other nations, have actually *reduced* their land area under cultivation. On the other hand, India and China, the most populous countries today, are already cultivating nearly *all* of their suitable land areas. The chief remaining biotic communities seriously threatened by agricultural expansion are the lowland coastal deserts found on most continents, and the tropical rain forests of the Amazon basin, southeast Asia, and Africa south of the Sahara.

Deserts can be very productive areas from an agricultural point of view. The chief limiting factor is the lack of fresh water, and it will remain limiting in these areas until desalinization of sea water becomes economically practical —probably around 1980.

Tropical rain forests (Fig. 7-2) present more of an ecological problem. There is no guarantee that it will *ever* be feasible to encourage large-scale agriculture there because **laterization** of the soil occurs when the forest is cleared. Despite the lush appearance of rain forests, their soils are usually thin and poor. When this soil is exposed to heavy rainfall by deforestation, silica and other soil minerals are leached downward and organic material is oxidized. If the thin topsoil is then eroded by the rain, a layer of aluminum and iron oxides is exposed, which in the air can form a hard, impermeable, red crust called **laterite.** Once formed, this crust appears to be relatively permanent and the vegetation that can grow there (including crop plants) is very sparse.

There is also considerable evidence that clearing large rain forest areas may raise temperatures and change climates. Feininger and other scientists working in northern South America have reported dramatic decreases in rainfall over large areas of the Amazon basin rain forest in Colombia during the past 26 years. The decreases have amounted to as much as 21 to 24 percent from earlier records of normal rainfall. A correlation exists between these changes and the fell-

Fig. 7-2. Clearing and burning tropical rain forest for cultivation in Costa Rica.

ing of vast tracts of rain forests in Colombia and neighboring countries. Feininger warns: "Should the decrease be widespread throughout the continent, the consequences could seriously retard the development of the tropical Latin American countries. . . ." *

* Tomas Feininger, "Less rain in Latin America," *Science* 160 (1968): 13–14.

Provided Latin American population levels are stabilized, many ecologists believe it may be best to follow the traditional Indian agricultural methods in Central and South America. These methods are usually called by the general term of **slash-and-burn agriculture.** The native typically clears a small area or areas of forest with axe and saw, allows the fallen logs to dry out, and then burns the plots. He plants a variety of crop plants, such as bananas, sugar cane, papayas, and cassava, in the same small field among the tree stumps. Several years later, when the nutrients in the soil are exhausted and erosion may be occurring, the farmer abandons his fields and moves on to clear another patch of forest and repeat the procedure. Meanwhile, secondary succession starts in his abandoned fields and eventually the young regenerating forest replenishes the soil nutrients. It still may take over a hundred years to restore a mature climax forest to the site, of course. This system will work as long as there is not an *excess number* of farmers operating in a particular area, increasing the probability of repeated use of the land before the forest has fully regenerated itself. Large banana plantations, exemplifying attempts at extensive clearing for agricultural purposes, have a high productivity in their first years, but not later. With prolonged use of the land for growing bananas alone, banana parasites (especially a root fungus) move in and destroy the plantings. Huge plantation holdings have been devastated by fungal parasites in Costa Rica, and the future of banana plantations in that famous banana-producing republic of Central America is in serious jeopardy. The only current way to fight the fungus, once established, is to abandon the fields and clear new areas some distance away; but this is a temporary holding action at best.

Defoliation and Bombing

Defoliant chemicals in considerable quantity are sprayed along roadsides and railroad tracks in this country to kill unwanted plant life. But nowhere in the world have herbicides been used in such quantities as in Vietnam during the late 1960s. In a massive landscape-altering program, approximately 13 to 14 million pounds of the defoliants 2, 4-D and 2, 4, 5-T (di- and trichlorophenoxyacetic acids), the *entire* production of these chemicals in the United States during 1967 and 1968, plus other compounds, have been sprayed on the Vietnamese forests by the U.S. military. The effect of these herbicides is to kill all vegetation from the ground up, rendering enemy units more visible in the thick natural forests and also destroying crops used as food. Agent Orange, a mixture of the n-butyl esters of 2,4-D and 2,4,5-T, was the most commonly used defoliant for general forest and brush clearing because it attacks a wide variety of broad-leafed plants. The C-123 sprayer aircraft can carry 1,000 gallons of the defoliant and spray their load in slightly over four minutes, flying just above the treetops at about 125 miles per hour (Fig. 7-3, top). Within several weeks, the forest is leafless and many or all of the trees may die, depending upon their susceptibility to the poison (Fig. 7-3, center).

Short-range changes in the sprayed biotic communities of Vietnam have already occurred. Nearly all flora and fauna are gone in the repeatedly sprayed areas. Forest fires have raged through the dead woods (living rain forests do not ordinarily burn), further exposing the soil to erosion and laterization. Near Saigon, a single forest fire destroyed nearly 100,000 acres of woods heavily treated with herbicides. Long-term changes are uncertain as yet, although mangrove swamps that had been heavily sprayed were still totally dead five years later. Mangrove trees help to stabilize coastlines and coastal waterways with their extensive root systems. Ecologists have noted dense bamboo stands, which prevent forest reproduction, developing in less heavily sprayed areas. Evidence also indicates an increase in the number of birth defects among babies of Vietnamese mothers who lived in sprayed areas.

Defoliation is not the only landscape alteration caused by warfare, of course. In Indochina, tremendous tonnages of bombs dropped on both forests and agricultural areas have resulted in craterization of the landscape (Fig. 7-3, bottom) on an immense scale, perhaps the largest excavation project

Fig. 7-3. Landscape alteration by defoliation and bombing in Vietnam. In the top photograph, C-123 are spraying Agent Orange defoliant (a mixture of 2,4-D and 2,4,5-T compounds) on a forest in central Vietnam. The center photograph shows a defoliated mangrove forest along the Saigon River; 100 percent of the trees have been killed by the application of Agent Orange. No regeneration has occurred in areas last sprayed over five years ago. The bottom scene shows B-52 bomb craters in agricultural country northwest of Saigon. Each crater is about 30 feet deep and 40 feet in diameter.

in mankind's history. The total cratered area throughout Indochina has been estimated to exceed the area of the state of Connecticut by 5,000 square miles. These craters are filled with water even late in the dry season, providing breeding sites for mosquitoes, ponds for aquatic snails that serve as intermediate hosts to parasitic worms that later infest man, and other pests. The total ecological and agricultural inpact of these bomb craters is yet to be measured, but it is likely to be great.

Urban Expansion

Today the worldwide phenomenon of urbanization (Fig. 7-4) is far more threatening to natural communities than is the expansion of agricultural clearing. The growth of urban areas will increase because of continued migration to cities, which increases the urban population, and high birth rates, which increase the total population. This increase in urban population, coupled with the trend towards spacious suburban residential development, is making continually greater demands on the available land around the urban fringe. In the United States alone, it is estimated that urban land area will increase from 21 million acres in 1960 to 45 million acres in the year 2000.

The increase in urbanized land is taken not only from surrounding terrestrial communities but also by filling in shallow marine bays. Land developers can purchase shallow areas of a bay and dredge up fill from adjoining bay bottom or bring in solid waste and soil fill from adjacent land areas. Prime examples of the dramatic extremes to which **dredge-and-fill** operations have been carried are seen in Tampa Bay in Florida, where 15 to 20 percent of the original water area has been filled in, and San Francisco Bay on the West Coast, with some 33 percent of the old bay filled. Such filling destroys principal spawning areas for a host of marine fish and invertebrates, destroys wintering places for migratory waterfowl, and may change the local weather patterns. San Francisco Bay is the most important stopping place for migratory waterfowl on the great Pacific Flyway. Approximately one million waterfowl winter on the Bay. If the filling trend continues at San Francisco (some 70 percent of the Bay is less than twelve feet deep and therefore easily fillable), these birds will lose their feeding and wintering grounds. Some meteorologists predict a sharp increase in summer temperatures (the cool summer fogs characteristic of San Francisco have already been noted to be less frequent now than formerly) and a drastic increase in smog. Not the least of the worries of San Franciscans who live or work in buildings constructed on

Fig. 7-4. Urbanization in a mountain valley of Peru: the city of Cuzco, 1970. Note the heavily terraced mountain slopes of the Andes behind the city; the people are using all available land either for cultivation or housing.

the unstable landfill should be the consequences of future earthquakes.

California and Florida are not ignoring the ecological problems of filling in bays. Many cities around bays have established so-called **bulkhead lines:** where no commercial filling and developing is allowed in a bay beyond a certain line (usually at the mean high water mark). Some counties have placed moratoriums on the sale and dredging of submerged lands, or passed laws requiring ecological surveys of bay areas before permits are granted for dredging, filling, or sale. The widespread public concern over the quality of the environment has been instrumental in bringing about these wise restrictions.

Clear-cutting of Forests

There are currently 182 million acres of national forests in the United States. These forest areas are administered by the U. S. Forest Service, an agency of the Department of Agriculture. Under a long-standing policy, formalized by a 1960 federal law, the 154 national forests are supposed to be maintained for multiple use, including timber, recreation, grazing, wildlife preservation, and watershed protection. In 1971, as a result of lobbying pressure from major western lumber companies, the President ordered the Forest Service to give top priority to private logging on this public land. To increase the immediate yield to the lumber companies, the Forest Service has increasingly authorized **clear-cutting,** an economical and efficient (100 percent yield) logging method that involves cutting of *all* trees in an area (the yield from selective cutting of mature trees is about 60 percent). Aside from the loss to recreation, grazing, and wildlife protection, perhaps the most dangerous result of clear-cutting (Fig. 7-5) is watershed degradation. A **watershed** is the area whose rainfall collects and drains downhill into a particular stream. The extent and character of the hillside cover of plant growth determines how fast the runoff is and how much sediment is picked up

Fig. 7-5. The ecological results of clear-cutting in a northwestern forest.

en route to the stream. Clear-cutting can cause a 7,000-fold increase in stream sedimentation and ruin breeding grounds for fish. Heavy snowmelt or rainfall on clear-cut slopes can cause rapid runoff and devastating downstream floods. These adverse effects of clear-cutting are of far greater economic cost to the whole human population of a region than is the expense incurred by lumber companies restricted to selective logging.

Public Works Projects

One of the major persistent threats to the natural environment is the giant public works project; ostensibly undertaken to "control" the environment and bend it to man's immediate needs, such projects often result in deleterious side effects that make the overall scheme an economic as well as an ecological loss.

In 1969 and early 1970 there was great controversy over plans to build a 39-square-mile jetport in Big Cypress Swamp, just north of the Everglades in south Florida. Not recognized by the planners was the ecology of surface water flow across the southern half of the state. This movement of water would be seriously disrupted by the runway construction for a jetport and associated commercial developments. Since the Big Cypress Swamp provides 38 percent of the water flowing south into Everglades National Park and the park is already suffering ecologically from years of drought in south Florida, the lack of water would irreparably destroy the basic ecology of this unique national preserve. Other powerful arguments, such as those concerned with expected air pollution and noise pollution, combined with this critical question of water supply, resulted in the defeat of this project by the aroused people of Florida.

Projects that may be desirable at one time in history are sometimes not accomplished then, but linger on in the minds of special interest groups and resurface later. A classic example is the U. S. Army Corps of Engineers' Cross-Florida Barge Canal. This mammoth project to cut a canal across the upper part of the Florida peninsula was proposed during World War II, with the intent of allowing barge and other coastal maritime traffic between the Atlantic Ocean and the Gulf of Mexico to avoid the threat of German submarines off southern Florida in the Caribbean. The war ended before the project could be initiated. In the late 1960s, the plan was revived and construction was begun by the Corps of Engineers, which also undertook an elaborate propaganda campaign, citing supposed economic benefits to be derived from cheaper shipping costs. Ecologists and conservationists soon discovered faults in the grossly favorable estimates used by the Corps, and challenged its supposed justifications and conclusions. Not only was the canal shown to be a potential disaster economically, but its proponents also failed to justify its construction in the face of the ecological damage it would have done. The proposed route would have created lengthy, biologically sterile channels in place of ecologically rich rivers, and destroyed more than sixty miles of natural communities adjacent to the canal, five major freshwater springs, and several large and unspoiled lakes (Fig. 7-6). The Army Corps of Engineers project was at last laid to rest by the personal intervention of President Nixon.

One of the largest public works projects ever conceived is the proposed Rampart Canyon Dam on the Yukon River in Alaska. The reservoir created by this proposed dam would flood 10,500 square miles and obliterate 400 miles of the Yukon River in addition to more than 12,000 miles of its tributaries and some 36,000 lakes and ponds of the Yukon Flats. Instead of the diverse natural communities of the area, the dam would substitute a huge and relatively sterile lake—useless as a breeding site for the 1.5 million ducks and 12,500 geese that currently reproduce in this valley. These ducks and geese, of course, fly south each fall to the main area of the United States, where thousands of sportsmen and amateur

Fig. 7-6. The River Styx, a cypress swamp near the Oklawaha River—former proposed site of the Cross-Florida Barge Canal project.

naturalists enjoy hunting them with gun or binoculars. More than 10,000 moose and other mammals such as caribou would have their habitat destroyed, a matter of considerable concern to the native peoples of the region, who depend on these animals for food. A salmon run along the Yukon that provides 800,000 fish per year for the local inhabitants would be lost. The sole purpose of the proposed dam is to generate huge quantities of electric power, which is not needed now and will not be needed in the next fifty years in Alaska. There are many other still unused power sources nearer the population centers of Alaska. Economically and ecologically, this proposed dam would be a disaster. But only an aroused and ecologically informed public can prevent a small body of narrow-sighted "development" or "dam-building" proponents from achieving such plans as these.

Summary

Landscape alteration may occur principally by fire, agricultural clearing, defoliation, urban expansion, clear-cutting, and public works projects. Periodic fire is a natural phenomenon that is actually necessary for the maintenance of certain climax communities. Agricultural clearing has become less of a threat to the environment today than in the past, although its expansion into the lowland coastal deserts and tropical rain forests of the world is anticipated. Defoliation caused by chemical sprays has caused severe ecological damage in Vietnam. Urban expansion, including the filling of bays, is destroying many natural communities. Clear-cutting of national forests is currently advocated by the lumber industry to increase timber yields. Giant public works projects that would harm great areas of our remaining natural environment are being constantly proposed. Prevention of environmentally harmful practices is usually accomplished only when the public protests with sound ecological and economic arguments in favor of an alternate course.

8
POLLUTION

Pollution of the environment simply means the introduction of an undesirable change in the particular features of concern, such as the constitution and quality of water, air, or soil. To a certain degree, the biological activities of all organisms cause some waste pollution in the environment, but man with his modern technology has increased his contribution by many orders of magnitude. In the natural world, decay organisms in the soil and water and the natural flushing action of massive water and air movements prevent the accumulation of organic or inorganic pollutants. When such cycles are upset either by a greatly increased quantity of pollution or by alteration of part of the natural restoration process, we have a major pollution problem, which often comes to our attention in unpleasant fashion.

The subject of pollution is almost endless and many excellent books have been written on its numerous facets. Some of these are listed in the Additional Reading section at the end of this book, for readers who are interested in studying this topic in more depth. Here we shall limit our consideration to the major types of pollution, with examples of some of the complexities and ecological factors involved.

Water Pollution

We seldom stop to consider the routine filtration and chemical treatment of urban water supplies today, but without this prophylaxis, disease organisms of many varieties (e.g., those that cause typhoid, paratyphoid, cholera, and various dysenteries) would be widespread pollutants among our human population. Pollutant materials or effects of other sorts are of much more concern than these disease organisms in today's society because we have had less success as yet in solving their particular pollution problems.

Change in the desirable characteristics of a water supply can occur in many ways. **Thermal pollution,** for example, is caused by the disposal of excess heated water into a body of water. This is usually done along rivers by power-generating plants and industrial factories that use water as a coolant. But in the construction of many large building complexes, also, the "solution" to the problem of hot water generated in the air conditioning system is to dump it into a nearby river or lake. Since most aquatic organisms have a body temperature at or near the temperature of the surrounding water and have a narrow range of temperature tolerance, the direct effect of a temperature rise is to change the rate of metabolic activity in using up food energy, which results in death. Since a rise in temperature of 18° F is sufficient to double the rate of many chemical reactions in plant and animal cells, it is easy to see why even a relatively small amount of thermal pollution is enough to cause lethal disruption in the organization of aquatic communities.

Storm drains often carry pollutants washed off paved streets directly into a water supply with no filtering or other treatment to which sewage is subjected. A major source of such pollution in our northern states is the salt used for de-icing streets during the winter. In the 1969–70 winter 77,000 metric *tons* of salt were used for de-icing in the densely populated Irondequoit Bay drainage basin near Rochester, New York. Investigators found that about 32,000 metric tons of salt went into the bay and the rest remained stored in the soil and ground water of the basin. The chloride concentration in this small bay (1 by 3.7 miles, with a maximum depth of 75 feet) has increased at least fivefold during the past two decades. The increase has been greatest at the bottom of the bay, and this has prevented complete vertical mixing of the bay's waters. The resultant lack of oxygen in the deeper waters could be expected to cause considerable reduction of the aquatic life in such areas (see page 101). The changing concentration of dissolved chloride could upset the internal salt balance of freshwater fish in the bay and the adjacent south side of Lake Ontario. During parts of the year the chloride levels exceed the U. S. Public Health Service recommended limit for human consumption. Since certain chlorine compounds are known to cause genetic mutation*, an increase of chlorine in water or food consumed by humans must be viewed with concern. Finally, the fraction of de-icing salt that is accumulating in the ground water may pose serious future problems for well water supplies in Rochester and other areas with similar de-icing programs.

Oil pollution has been a major public concern since the day in March 1967 when the giant tanker *Torrey Canyon* hit a reef off the Cornish coast of England and broke apart, releasing 117,000 *tons* of crude oil into the sea. Thousands of tons of oil drifted ashore in England and in France, on the other side of the English Channel. When the British found that the oil on the water could not be ignited, they sprayed some 12,500 *tons* of detergents on the oil spill to try to emulsify it (i.e., break the continuous oil sheet into tiny globules) and on the oil-covered coastal rocks to remove the oil that had already reached the shoreline. This course was not very successful either at sea or along the coast.

Ecologists investigating this disaster found that the chief immediate biological effect of the floating oil was to kill marine birds. When they landed on the surface or dived into the water for prey, the oil coated their feathers and destroyed the birds' ability to float. The death counts in two species alone were at least 20,000 murres (penguin-like birds) and 5,000 razorbills. The oil did not kill the intertidal zone life such as limpets and barnacles, or much of the plankton in the open sea. Detergents, on the other hand, turned out to be extremely toxic to the marine plankton and to the intertidal forms living on and around the English coastal rocks. The French did not use detergents; instead, they dropped powdered chalk on the floating oil. The chalk-oil complex sank to

* A genetic mutation is a change caused by a chemical, radiation, or other agent, in the hereditary material composing the chromosomes. It will be inherited by later generations and is usually harmful in effect.

the ocean bottom, preventing much of the oil from drifting into shallow coastal waters where it would do more harm to abundant marine life. The ecological lessons learned from fighting the Torrey Canyon oil spill have proved valuable in handling later spills from ships and sea oil-well platforms, and in passing federal and state legislation to prevent them.

Vessel pollution, involving discharge of wastes from ships into oceans, lakes, reservoirs, and rivers, is becoming an increasingly significant problem with growth in the number of merchant ships and the boom in recreational boating in the United States. **Dredging** operations in estuaries to improve ship channels or produce landfill create **silt pollution** on a grand scale.

Irrigation pollution occurs when agricultural water leaches out minerals or salts from irrigated fields (Fig. 8-1) and the runoff carries such pollutants to the nearest stream or lake. The **agricultural runoff** may also carry excess nitrates or other chemicals present in fertilizers, or in herbicides and insecticides from land being treated with persistent pesticides (a topic to be considered in a later section). In areas of extensive mining such as West Virginia, mine drainage can cause very harmful pollution in adjacent water supplies. Streams

Fig. 8-1. Irrigation in an agricultural field in Florida.

flowing out of mine shafts and runoff water that crosses slopes of rock removed from mines often become excessively acidic from dissolved chemicals.

Still another polluting by-product of our modern technology is **radioactive waste.** Perhaps the prime sources of such pollutants are research laboratories and hospitals that use radioisotopes extensively for tracing the metabolic pathway of materials in organisms and for a variety of diagnostic purposes in humans (localizing tumors, determining defects in normal circulatory anatomy, and determining blood volume). After passage through a patient, a waste radioisotope solution is still radioactive. Weekly collection and burial of this material in lead-lined containers is not generally feasible because of the need for special toilets, special handling, and other expensive requirements. The solutions containing these radioisotopes are merely flushed down the drain and thus enter the general sewer system, ending up eventually in some body of water. Though potentially hazardous in the future, radioactive waste from all sources today seems minor compared to other forms of pollution freely released into the environment.

Finally, one of the most common and currently controversial forms of water pollution is the process of **eutrophication.** This is the promotion of plant and animal growth in an aquatic ecosystem by adding substantial amounts of nutrients to the water. When eutrophication is accelerated by man's activities, pollution is rapid, for small unicellular algae and other aquatic plants have a phenomenal rate of increase under advantageous conditions, and can use up most of the free oxygen present in an overnourished body of water. Larger animals like fish then die from lack of oxygen. Three principal sources of excessive supplementary nutrients are runoff from stockyards, runoff from heavily fertilized farmland, and outflow from sewage plants that have incomplete sewage treatment and do not reclaim nutrient chemicals. A controversial source is many laundry detergents, which contain phosphates. Environmental scientists have worked to ban the growth-promoting phosphates from detergents, although some sanitary engineers and the major soap manufacturers claim that most commercial organic products (e.g., hand soaps, cleansers) contain phosphates anyway and that the removal of detergent phosphates alone would be useless. The chemicals currently used in non-phosphate detergents present a health hazard to children in the home and to adult users who may be sensitive to them. We have yet to strike the proper ecological balance in the inclusion or exclusion of phosphates in detergents, although their pollution-promoting tendency through eutrophication is well documented.

Air Pollution

Components

Gaseous wastes and particulate matter such as soot are the principal components of air pollution, but a surprising variety of substances pollute our atmosphere, especially around developed urban areas and manufacturing sites. Chemical pollutants present in significant quantities include carbon monoxide, a colorless and odorless gas. Carbon monoxide combines readily with hemoglobin in the red blood cells, and thus prevents the hemoglobin from carrying oxygen. Death from oxygen deficiency occurs if a person is exposed even to low concentrations of this gas. Sulfur oxides (especially sulfur dioxide), nitrogen oxides, and hydrocarbons are the principal components of "smog." Pollutants present in minor quantities include ozone and other oxidants, arsenic, asbestos, beryllium, fluorides, lead, organic and inorganic pesticides, and sulfides.

Sources of Air Pollution

In nature, air pollution arises principally from forest fires and volcanic eruptions. Generally, the infrequency with which these events happen and the short time they last insure that pollutants released into the atmosphere are soon dissipated by air movements. Since the advent of man as a technological animal, however, air pollution has become a much greater and more persistent feature of our world (Fig. 8-2). Today, the principal sources are processing industries and

Fig. 8-2. Air pollution over the city of Los Angeles (top) and spreading ninety miles east of Los Angeles up to 5,000 feet elevation in the mountains (bottom).

power-generating plants that burn coal and oil, and the internal combustion engine. Some cities have passed ordinances limiting the allowable amount of pollutants to be emitted from industrial plants, and state and federal standards are being established. An especially concerted effort by environmentalists, government, and automobile manufacturers has been made to reduce the emissions of the automobile internal combustion engine to an acceptable level. The ultimate solution to this vexing problem would be to switch to an alternative type of engine with less potential for pollution, one that uses steam or electric power instead of burning gasoline. Since California and other states have set deadlines in the mid-1970s for meeting minimal emission standards, Detroit technology is working particularly hard on this problem.

Environmentalists have been less successful in achieving the regulation of emissions by huge power plants. In the Four Corners area in the southwestern United States, where the borders of Arizona, New Mexico, Utah, and Colorado meet, a series of large power-generating plants has recently been built. The enormous amounts of air pollution released daily from these coal-burning plants are currently covering thousands of square miles. Conservationist groups and the Navajo Indian tribal council are attempting legal action to bring about the abatement of this air pollution.

Effects of Air Pollution

The principal known effects of air pollution are, of course, on human health. In cities with a high incidence of air pollution, human longevity is decreased through a notably higher frequency of emphysema and other lung diseases. Of course effects are not restricted to people. Many species of flowers, shrubs, and trees are unable to thrive or even survive in cities such as Los Angeles. Commercial and amateur horticulturists in southern California who raised such flowers as orchids and bromeliads have seen disaster strike their greenhouses through smog damage to both blossoms and leaves, and have been forced to move or abandon their floral enterprises and avocations.

The damage from air pollution is not restricted to the immediate urban or factory area of the source. Smog moving out of the Los Angeles basin is damaging agricultural crops and killing mountain pine forests as far as 100 miles away. In late 1970 an aerial survey of the Angeles National Forest, which borders Los Angeles, showed that 261,000 Ponderosa pines had been damaged by smog. Not only are the pine community food web and other biological relationships being altered; the recreational and lumbering activities possible in such areas will undoubtedly change with the drastic rate of loss of the dominant tree species.

Air pollution may also cause major changes in weather and climate. The town of La Porte, Indiana, on the eastern side of the tip of Lake Michigan, has had an excessive run of bad weather for the last forty-five years. Total precipitation has increased more than 30 percent in forty years, while the number of hailstorms, days with hail, and total precipitation at La Porte greatly exceeds the records of nearby weather stations. The growth of air pollution in the Chicago-Gary industrial complex thirty miles west of La Porte is closely correlated with these trends; even the days of bad weather in La Porte usually correspond with days of bad air pollution in Chicago. The quantities of heat and water vapor in the pollution emitted by the Chicago factories apparently cause the formation of cumulus clouds, which then are blown across the tip of Lake Michigan, picking up additional moisture. At about the time these clouds reach La Porte, the particulate matter in the mixture of pollutants provides nuclei for condensation, and the result is precipitation.

The problem of air pollution is not confined to the outside environment, however. Smoking in closed areas like classrooms, conference rooms, airplane cabins, or cars causes high levels of particulate pollutants and gases such as carbon monoxide. In a January 1972 report to Congress, the U. S. Surgeon General announced evidence that the health hazard from such pollutants is not limited to smokers. Nonsmokers in a smoke-filled room may be exposed to harmful levels of carbon monoxide that are especially threatening to persons already suffering from chronic bronchopulmonary and coronary diseases. Carbon monoxide in such concentrations (20 to 80 parts per million) adversely affects human sight and hearing and hampers an auto driver's ability to respond to light and to judge distance. Higher concentrations of carbon monoxide cause death by interfering with the transport of oxygen in the blood. Of course, the smoker himself is subjected to far greater pollutant levels than is the nonsmoker. Vincent Schaefer stated in a 1969 article in the journal *BioScience* that "in the process of smoking, the individual insults his lungs with a concentration of at least ten million smoke particles per cubic centimeter. This is a concentration that is 10 to 100 times greater than is encountered in a very badly polluted urban area like Los Angeles or New York City." *

Solid Waste Pollution

Solid wastes (Fig. 8-3) in the United States are commonly disposed of by one of three methods: incineration, land-filling, or reclamation. **Incineration,** or burning of solid matter, is now considered undesirable because of its major contribution to air pollution. Although burning is still the principal method of disposal in some sections of the country, it is increasingly restricted to sparsely populated areas. **Land-filling** or **sanitary landfill** disposes of solid wastes by dumping them into a pit, natural canyon, marsh, or coastal estuary. As the depression is filled, the waste material is covered with soil. This is the commonest type of garbage and waste disposal system for towns and cities, because of its low cost, lack of processing trouble, and avoidance of air pollution. The city of Charleston, South Carolina, found that it cost about 65 cents a ton to dispose of garbage by using the landfill method in a marshy area on the coast. After clay was brought in to put on top of a 25-foot layer of waste, the land could be sold for other developments. Unfortunately, some coastal cities still find it eas-

* Vincent J. Schaefer, "Some effects of air pollution on our environment," *BioScience* 19 (1969): 896–897.

Fig. 8-3. Solid waste pollution outside one of our major western cities.

ier to dump their garbage and trash at sea, thus causing water pollution. This marine "disposal," with large-scale consequences for *all* areas around a sea, cannot be considered socially or ecologically responsible.

Reclamation is the environmentally beneficial way to avoid solid waste pollution and is rapidly gaining favor as new technological methods are developed to make it competitive economically with landfill. Reclamation is simply the reclaiming and recycling of waste materials. To take a single example, the largest reclamation problem in the United States is paper. Unfortunately, 50 percent of this paper waste is cardboard, and only 20 percent of the cardboard is recyclable under present methods. Among the more interesting methods of recycling, Atomic Energy Commission researchers have fed high-quality paper (onionskin and bone paper) experimentally to cattle; the cattle, whose digestive systems contain microbes capable of digesting cellulose, grew fat on the diet. Paper waste could also be used to grow yeast cultures, as for breweries. The major recycling of paper waste, though, is channeled into making new paper products.

Tin-coated steel cans used to be the commonest food storage container on the supermarket shelf; today aluminum cans, which are easily reclaimed, have captured a sizable percentage of the can industry's production. Unfortunately, all that glitters is not gold; nor will it rust. The cans that are not brought back for recycling but lie abandoned on roadsides are much more rust-resistant than the old-fashioned steel can. Steel cans now have resin coatings instead of tin; the resin can easily be removed during re-smelting, making it feasible to recycle steel cans. Because of their degradable properties if abandoned in nature, the new recyclable steel cans may be ecologically preferable to the less expensive aluminum cans.

Glass containers can obviously be directly recycled through re-use. Other imaginative ways of using glass include grinding it and using the granular material in insulation, vinyl floor tiles, building walls, and road surfaces. Plastics are one of the most challenging items to reclaim because they cannot be broken down in nature by biological action or weathering. They can be melted, but the resultant sticky mess is not readily disposable or recyclable. Fortunately, it has been found that specific impurities can be added to cause plastics to break down after a certain period of time. Manufacturers could adjust the composition of plastics for various articles according to the expected length of service of the particular item.

Finally, we might mention one dramatic solution to the problem of disposing of solid wastes. A former head of the Atomic Energy Commission, Glenn Seaborg, suggested breaking garbage into individual atoms with a hydrogen fusion torch. Unfortunately, the amount of resulting radiation would appear to make this idea impractical at present.

Pesticides and Pollution Problems

The term **pesticide** refers to any chemical substance used for the regulation of population growth in a species regarded as a "pest." Pesticides include **herbicides,** used for the control of herbaceous plants; **insecticides,** used for insect control; and **biocides,** general pesticides that may be used to kill all life. The principal uses of pesticides have been to kill plant or animal pests, to stop the spread of disease, and to protect crops, all perfectly laudable aims. Since the end of World War II, however, major problems have arisen from the development and overuse of certain pesticides. These have centered upon the increasing use of indestructible pesticides; their accumulation in the ecosystems of the world; increasing pest resistance to them; and their potential danger to humans and organisms in general.

Until World War II, the problem of persistent molecular types of pesticides did not exist. Before that time, naturally occurring compounds such as nicotine (extracted from tobacco) and lead arsenate were used as pesticides. But with the advance of chemical knowledge, chemists began to synthesize pesticide compounds *not* derived from natural sources. Among these were the chlorinated hydrocarbons, which include such insecticides as DDT (dichlorodiphenyltrichloroethane) and the DDT derivatives, DDE and DDD; dieldrin; chlordane; and the herbicides 2, 4-D and 2, 4, 5-T. Unlike naturally derived pesticides, these compounds are not broken down by biological action and they persist for many years in an unchanged state. Thus it was found that when DDT was sprayed on walls of houses and other living areas, malaria-carrying mosquitoes died even if they landed on the surface months after spraying. DDT aided in conquering malaria throughout the world and prevented typhus epidemics during World War II as well. Its use on agricultural crops (Fig. 8-4) seemed to promise untold benefits by increasing production.

Fig. 8-4. Spraying of pesticides on crops to kill specific pests has been instrumental in increasing modern agricultural production, but excess chemicals that drift in the air to adjacent wild areas can cause considerable harm to non-target organisms.

Then in 1946 it was first noticed that several fruit fly strains had become resistant to DDT. By 1948 eleven more species of flies had become tolerant. In 1958 seventy-eight additional insect species were resistant, and in 1969 at least 224 species of insects were found to have developed effective resistance to the lethal effect of DDT. This increasing insect resistance has become a classic example of rapid evolution observed in progress. In brief, the widespread use of DDT caused strong natural selection for genetic strains in an insect species that contain genes producing detoxifying enzymes, and against those genetic strains susceptible to DDT. There was also strong selection for pesticide-avoidance behavior, where genetic strains with behavioral patterns that kept the insect away from a pesticide tended to survive and reproduce more than those who caught full exposure to the chemical. In several tropical species of *Aedes* and *Culex* mosquitoes, the populations that in the 1940s and earlier were primarily house feeders on sleeping humans now will bite only outside houses, indicating that genotypes producing this behavioral pattern have survived to replace those that used to enter houses, alight on a DDT-sprayed wall, and die.

The Peruvian cotton disaster in the Cañete Valley in the 1950s clearly pointed up the result of the spraying that caused rapid selection of resistant insect genotypes. In this Peruvian coastal valley, major cotton production from its start in the 1920s until the late 1940s was aided by old-fashioned insecticides such as nicotine sulfate and calcium arsenate. The limited pest insect complex in the valley was kept under fair control. Then growers decided to employ the newest synthetic organic insecticides, including DDT, benzene hexachloride (BHC), and toxaphene. Excellent increases in cotton yield resulted as the modern pesticides appeared to decimate the insect populations. The new pesticides blanketed the valley. But after several years, some of the old pests had developed resistance to the insecticides and had become increasingly destructive. New pests appeared. The pests were immune to the organochlorine insecticides. By the 1955–56 season, the sprays had been increased greatly but insect resistance was general and the crop was now decimated, the yield being one of the lowest ever recorded for the area. To survive, the cotton growers had to abandon these insecticides and begin an integrated control program, using the older non-synthetic chemicals and bringing back natural biological control of the pests by encouraging their native enemies. Under this program, cotton yields increased greatly, the old pests decreased to their old levels, and the secondary pest outbreaks of the new insect species that had moved in soon disappeared.

If persistent pesticides were confined only to agricultural areas in developed nations, the matter would be important from an economic viewpoint and as an example of man's upsetting the natural balance to his own detriment, but it would not affect most of us directly. Unfortunately, persistent pesticides have two nasty features: they stay around for a long time and are easily spread. Aerial spraying results in such extensive vaporization that in tests over corn fields as little as 5 percent of the pesticide spray actually reaches the corn field. The rest drifts through the lower atmosphere to adjacent areas; even part of the pesticide residues in the sprayed field will wash off the leaves in the next rain and be carried out of the field. When pesticides reach a stream or river, whether by aerial drift or by rain runoff, they are usually absorbed into suspended dirt particles and carried by water movement to the oceans. In 1967, DDT was discovered in the body tissues of penguins and other wildlife in Antarctica, indicating that DDT was dispersing generally throughout the world, even to areas where no spraying had ever been done.

The persistent nature of molecules of DDT and its chlorinated hydrocarbon relatives, and their affinity for body fat, cause them to accumulate in the upper levels of natural food chains in all parts of the world. In the typical food-pyramid relationship, animals at each trophic level eat large numbers of organisms from lower levels. If these prey have any DDT in their tissues, it will be incorporated in the body tissues of the animals at the next level, and so on to the top carnivores. Hence birds of prey, such as the bald eagle or osprey, being at the top of their respective food chains, should receive the highest concentrations of DDT. As it turned out, this is exactly what happened; eagle and osprey populations began declining in the 1960s with corresponding increases in DDT concentrations in their bodies, and it was soon discovered that the reproductive success of these species had suddenly become almost zero. The cause was thin fragile eggshells, which easily broke under the parent's weight. In subsequent laboratory experiments DDT was found to interfere with hormonal balance, lowering the calcium ion content in the female bird's blood, which in turn lowered the calcium available for eggshell production. Since the initial discovery that DDT causes eggshell breakage in ospreys and bald eagles, similar devastating losses have been reported in populations of robins, brown pelicans, and many other birds.

Mammals as diverse as sea lions and bats have also shown dramatic reduction in reproductive capabilities in recent years. In Arizona, for instance, the populations of several of the 28 recorded bat species have decreased drastically. Between 1963 and 1969 one colony of Guano bats (*Tadarida brasilensis*) near the town of Morenci decreased from over 25 million individuals to only 30,000 individuals. These bats fly over agricultural fields at night to feed on flying insects, and their annual migratory route takes them over agriculturally developed areas of Sinaloa and Sonora, Mexico. With their long life spans (to 21 years), bats are particularly vulnerable to accumulating a lethal load of chlorinated hydrocarbons from the pest insects they eat. Large numbers of dead Guano bats, containing DDT and DDE in body fat, were found around the feeding range of the Morenci colony. E. L. Cockrum, a biologist at the University of Arizona who investigated this problem, has stated, "While the future of the southwestern Guano bats seems bleak and while the cause of the crisis, insecticide poisoning, cannot be proved beyond all doubt, one outcome can be agreed upon. Bats of the great Morenci maternity colony no longer destroy, as they once did, *40 tons* of insects nightly." *

Of course man, being a top predator in many food chains, has not escaped the flow of DDT through the world's ecosystems. Even Eskimos have been found to have 3.0 parts of DDT per million parts of tissue in their body fat, while in

* E. Lendell Cockrum, "Insecticides and Guano bats," *Journal of the Arizona Academy of Science*, vol. 5, no. 4 (1969), editorial page.

areas such as Hungary and Israel, with high use of chlorinated hydrocarbon insecticides, the concentrations in human tissue are higher (12.4 ppm and 19.2 ppm, respectively). In the U. S. population, 5 to 20 ppm is the common range of DDT concentration. Unfortunately, it reaches its highest concentration in mother's milk, and such milk may actually be unfit for infant consumption. DDT has also been shown to interfere with sex hormone balance (it mimics the action of the female hormone estrogen) and cause infertility in female rats. The same pattern of infertility caused by hormonal blockage of ovulation was observed in American women and reported by University of Washington scientists in April 1971.

Biological Control

To avoid the pollution problems inherent in the use of persistent pesticides, economic entomologists and ecologists have suggested the use of organic phosphate compounds (e.g., parathion, malathion) or other biodegradable chemicals, combined with biological control methods.

Biological control is simply the suppression of reproduction of one type of organism by utilizing some feature of its biology or physiology to destroy it or by use of another organism. The oldest known use of biological control dates to several thousand years ago, when the Chinese placed bamboo poles between orchard trees to allow predatory ants to raid pest insects easily.

The principal biological control agencies used today are parasites, predators, and microbial diseases. Natural parasites of pest insects or plants, such as parasitic wasps, may be mass-cultured in the laboratory and then released as needed in crop areas. Predators such as ladybug beetles are sold by the thousands (1 gallon at $6.50, equal to 125,000 beetles!) to control scale insects and aphids. The most effective control is offered by bacteria and viruses, though. These are species-specific, killing only one particular pest, and do not accumulate in the environment like chemicals, merely joining the already existing microbe population. Thus far, insects have apparently failed to evolve resistance to these useful strains of microbial disease. The most effective biological agent in the Japanese beetle control program of the U. S. Department of Agriculture in the eastern United States is a viral disease that attacks the beetle grub underground.

The principal *disadvantages* of using microbes as biological control agents are fourfold.

1. Because of the specificity of most interspecific interactions of these types, one agent microbe is usually needed for each pest species.
2. Chemicals act much faster than microbial disease; so further damage may be done or movement to adjacent fields may occur after the microbes are released but before they are effective.
3. The microbes must be cultured in the laboratory in the same insect species that is acting as a pest in the field; so one must culture the host as well as the microbe.
4. Because of Food and Drug Administration limitations, there is not a single viral control organism being sold commercially for use by the general public on agricultural or horticultural pests. This last difficulty makes the earlier disadvantages almost academic, though we may anticipate some relaxation of the F. D. A. rulings in the future as the use of biological control methods becomes more imperative.

Other biological control agencies that are useful alternatives to pollution-producing pesticide chemicals are available. **Resistant varieties** of crop plants are being developed by plant geneticists who select variant genotypes with chemical defenses such as alkaloids or mechanical protection like thicker bark. **Male sterilization** has been a successful technique to eradicate or control the screwworm fly and several injurious fruit fly species. Male flies reared in large numbers in the laboratory are sterilized by radiation and released into wild populations weekly. If a sterile male mates with a virgin female, she will lay her eggs and they will fail to develop. The screwworm fly has been eradicated from cattle-raising areas on Caribbean islands and in Florida by such mass-release programs. **Organic attractant chemicals** that influence the chemo-sensory behavior of pest insects, such as sex attractants, can be synthesized in large quantities and used to attract and destroy large numbers of adults or at least confuse the sensory environment and so prevent matings. **Specific hormonal insecticides,** which upset hormonal balance within an insect when absorbed through its cuticle and cause reproductive failure or death, do not affect other species in the same area. Finally, **radiation signals** can be utilized to attract and destroy insects or to upset the biological clocks and seasonal cycles of pest species. Most night-flying adult insects are attracted to ultraviolet light, and electric grids or other trap devices coupled with ultraviolet light can be effective control devices for some species. One can change the photoperiod over truck gardens (valuable vegetable crop land near cities) with short periods of artificial lighting at night and make insect pests go into diapause or prevent normal egg-laying patterns.

The limitations of biological control as opposed to chemical control are its greater cost, more complex treatment procedure, and occasional uncontrolled reproduction of the biological agent. The generally greater cost arises from rearing the parasites, sterilized males, or other control agents in a laboratory. The man-hours involved in caring for the living organisms under tightly controlled conditions usually far exceed the manufacturing time required for a chemical product that could treat the same agricultural field. If repeated treatments with living material are needed in the field area, the complexity of operation in rearing and releasing biological control agents is considerably greater than spraying a pesticide from the air or ground. Occasionally control over the parasite or pathogen species is lost and it jumps to other, "good" species in the fauna of a treated area. This is more likely to happen where the agent species does not occur naturally, and it can occur even though the agent is species-specific on the pest species alone in other areas. We should also remember that many of the non-persistent, biodegradable pesticides may safely be used on crops. Thus the best control plan, which avoids the limitations of pollution-producing chemicals and the occasional lapses of biological control alone, is an integrated control program, using several control methods simultaneously in conjunction with a thorough study of the vulnerable periods in the pest's life cycle. The success of such a program in the cotton fields of the Cañete Valley of Peru after the disaster with chlorinated hydrocarbons is indicative of the effectiveness of this concept.

Summary

Pollution of the environment is simply the introduction of an undesirable change in a particular feature of concern to man or other organisms. Among current major problems in this area are water pollution (eutrophication, thermal changes, storm drainage, oil, vessel wastes, dredging, agricultural and irrigation runoff, and radioactive waste); air pollution; solid waste pollution; and pesticides of several types, especially the chlorinated hydrocarbon insecticides and herbicides. Biological control and integrated control programs are ways to handle pest problems without the use of persistent pesticides.

9 POPULATION GROWTH IN MAN

The Population Explosion: Nature of the Problem

The root of almost all the threats to natural communities and the ecology of our earth is the explosive growth of the human population. Man's burgeoning numbers are not restricted to the Asian and other less developed countries of the world. The United States now has some 208 million people, and crowding is becoming a fact of life even in such remote areas as Yellowstone and Yosemite National Parks. The inhabitants of the Los Angeles basin and of the New York metropolitan area have been familiar for years with the smog and other problems attending overpopulation in a small area. Now whole states, such as Oregon, have awakened to the dangers of unlimited population growth and are attempting to regulate or halt it. Special groups within the United Nations, as well as many private organizations such as the Population Reference Bureau, are attempting to diminish or reverse the precipitous worldwide rise in population.

Today's high growth rate is a relatively recent phenomenon, dating from about 1650 A.D., although there were several earlier surges in population growth. Man's primitive primate ancestors are believed to have been African forest species that lived alone or in small, loosely knit groups. With the development of bipedalism (walking on hind limbs alone), tool-making, and perhaps speech, early men began living on the open plains in family groups. About 600,000 years ago, Peking Man, Java Man, and similar *Homo erectus* forms moved into caves and other more permanent shelters, utilized fire, and made weapons. Man's survival rate was enhanced by these new cultural features, and this major **cultural revolution** set off the first real surge in world population growth.

Approximately 8,000 years ago, the second notable increase in rate of population growth occurred with the advent of the **agricultural revolution.** At this time in several major regions of the world man changed from a primitive, hunting-and-gathering level of existence involving every member of a nomadic group, to an agricultural society where only a part of the population could raise enough food to feed the whole society. A sedentary mode of existence, raising crops and livestock, presented fewer hazards to the average man and his family than the former migratory or wandering way of life. Food in the form of grains could be stored for times of scarcity and drought. The death rate consequently decreased and the live birth rate increased.

The third and greatest rise in man's population size began in the mid-seventeenth century and is still accelerating. This increase in growth rate was triggered by the **scientific-medical revolution** that started around 1650 A.D. Growth was further accelerated by social changes associated with industrialization. Medical knowledge and public health measures began lowering the death rate significantly at that time, and this trend has continued up to the present day. In highly developed countries such as Japan, the United States, and nations in Western Europe, the birth rate has also fallen, but not as drastically as the death rate. Thus their growth continues, though at a low rate in some countries. Technological advances in housing and transportation of food have made it possible for man to inhabit even remote polar or desert areas today, where previously population growth was limited by the lack of a sufficient natural supply of food or suitable living conditons.

In underdeveloped countries such as India and Pakistan, the medical and scientific advances that originated in Europe did not arrive until World War II. From 1940 to 1945, the Allies introduced public health practices such as malarial control and sanitation of public water supplies on a large scale, and the death rates in these countries fell. The high birth rates characteristic of these unindustrialized nations continued unabated. The gap between the birth rate and the death rate, then, has accelerated our population growth rate in all countries to the point where it is estimated that the world population will jump from our present 3.7 billion to 7.4 billion by the year 2000. The trend of the world's population growth since the time of Christ is shown in Figure 9-1.

Population control has, of course, been a problem for every species of animal or plant on this planet, but over time these populations maintain more or less constant population sizes below the carrying capacity of their environment. We have already looked at some of the regulating mechanisms involved (Chapter 5). These density-dependent and density-

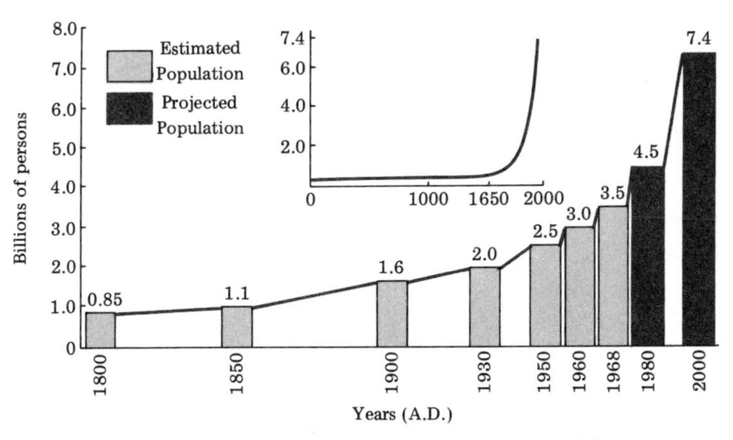

Fig. 9-1. The growth of the population of the world from 1 A.D. to 2000 A.D., with population estimates for the years 1800–2000 shown in detail.

independent factors have been found in all organisms whose ecology has been thoroughly investigated, and it would be peculiar indeed if they were lacking in one species, Man. Since the problem of a high rate of increase seems to be relatively recent in the history of man on earth, it is worth examining past and primitive societies to see how they were affected by population growth and how they controlled it. In this examination it will become evident why we have lost most of the natural or culturally instituted regulating factors that operated in primitive groups. Some were lost when civilization began in the agricultural revolution, and the great majority had disappeared by the time we entered the scientific-medical revolution.

Population Growth in Primitive Societies

The chief ecological and evolutionary facts that we found in considering population growth in plants and animals apply to man also: population stability is a prerequisite for survival, and selection operates against organisms in groups where population control is ineffective or absent. In primitive societies of the past and present, population stability has been essential for survival. Since a hunting and gathering group requires a large territory to furnish what it needs all year, a low population density in the form of widely spaced groups is a necessity. Should a group exceed the carrying capacity of its area, individuals could survive by voluntarily emigrating to a new area, if they were unopposed by a neighboring group and if suitable land were available. Excess individuals could also fail to produce offspring or be eliminated from the group involuntarily by reason of famine, disease, or warfare. These last three factors were considered by Thomas Robert Malthus, an eighteenth-century Englishman, to be instrumental factors in controlling population growth. His famous 1798 and 1803 essays on the principles of human population growth were the first to warn of the dangers of overpopulation and its probable consequences for Man. In his writings, primitive societies provided a valuable base of comparison with industrial societies of his time.

Since strong selection would presumably occur against groups of individuals that did not have some form of population control, it is not surprising that man himself added cultural traits to control reproduction in his societies. Cultural evolution has produced an amazing variety of such checks to reproductive excess, and space permits only a partial list. Most societies have extensive marriage and reproductive taboo systems. The consecration of virginity is common to almost all modern human groups as well as many primitive groups. There is often punishment for out-of-wedlock sexual experience or for early motherhood in societies permitting premarital sexual activity. Frequently, primitive societies have minimum and maximum ages set by custom for motherhood or fatherhood. Required child spacing is not uncommon. In societies with this custom, reproductive contact by the couple is prohibited or restricted until a new child reaches a certain age. Widows are often forbidden to remarry for a period.

In some societies, celibacy is advocated as a high calling. In certain South Pacific societies and in servant classes in ancient civilizations, castration of males insured that celibacy would be observed. Certain African and other tribal groups practiced infibulation of young females as a pubertal initiation rite. Of course, this surgical removal of the clitoris removed much of the sensual pleasure associated with intercourse and hence perhaps reduced its frequency for those females. Prostitution and homosexuality were utilized in diverse groups as forms of sexual conduct that tended to reduce the reproductive contribution of the participants.

A very common practice in primitive groups was exposure of infants or other forms of infanticide. Interestingly, the Biblical book of Exodus describes how the Egyptian pharaohs at the time of Moses (approximately 1550 to 1447 B.C.) used exposure of male infants as a population-control measure against their Hebrew slave populations. Infanticide was commonly practiced by the Greeks and Romans at the height of their civilizations, and it is probably universally characteristic of past and present-day primitive groups and rural cultures. Exposure of aged or sick persons who no longer contributed to the food-gathering capacity of the group was also a common practice among North American Indian tribes and other societies. Abortion, or killing of the fetus by mechanical and chemical means, was practiced by most primitive groups of peoples and is still common today, especially in rural peasant cultures.

Sometimes physical combat within and between groups for reasons unrelated to the normal warfare aim of acquiring territory played a significant role in limiting population size. The Jívaro Indians of eastern Ecuador, for example, spent much of their lives in hunting other Jívaro enemies for the non-territorial reasons of retaliation and revenge for injuries suffered in previous raids. Cannibalism in some groups was a form of warfare devoted partially to the aim of securing human food. Certain societies used war and raids to obtain enemy individuals for religious sacrifices. Records of the Aztec civilization of central Mexico show that human sacrifice could be a significant factor in adding to the death rate in a population. Human sacrifices were adopted by the Aztecs early in the fourteenth century and continued for about two hundred years until the conquest of Mexico by Cortés in 1520. Captives were sacrificed at almost every festival. On great occasions, such as the coronation of a king or the consecration of a new temple, appalling numbers of prisoners were slaughtered on the altars. The *yearly* sacrifices throughout the empire are estimated from excellent historical records to have ranged from twenty thousand to over fifty thousand people. At the dedication of the great temple of Huitzilopotchli in 1486, the ceremony included the sacrifice of some seventy thousand captives!

We might close this consideration of early population-regulating factors by mentioning that the practice of slavery has often limited the subject population in harsh ways. In the eighteenth and early nineteenth centuries, African slave traders and tribal raiding parties killed large numbers of Africans while on expeditions to secure slaves for export to areas such as America and the Caribbean. The exploiting groups of men in these and other periods of history had, of course, direct and absolute control over the later reproductive activities of those taken alive for slaves. In early civilizations as well as at least one twentieth-century society (India), slaves were sacrificially killed along with other members of the household when the head of a noble family was killed in battle or died of natural causes. The most common cultural rationales for this practice were that the departed soul needed companions to serve him in the next world, or that members of the household could hardly desire to continue living when their titular head was dead.

All these factors aided the three main Malthusian control mechanisms of famine, pestilence, and territorial warfare in limiting the size of the primitive human society. It is notable that the Malthusian factors are beyond the control of the individual and often beyond the control of his society. The culturally imposed factors that we have reviewed in these past few paragraphs were all voluntarily imposed by the societies themselves (or at least by segments of them). While often instituted ostensibly for other reasons, primarily related to mystical or nature-worship religious systems, these cultural factors exerted effective population control.

The great majority of these culturally instituted control measures are no longer morally acceptable to modern societies, and it is not likely that mankind will revert to the primitive type of religious structure in which they had their basis. Except for their historical interest, most do not offer any serious help for man's present problems. The chief exceptions are the social emphasis on female virginity (still current in our society), non-reproduction outside of marriage, and advancing the minimum ages for sexual reproduction within marriage. These could all contribute significantly to effective population control, in combination with other measures.

To summarize what we have learned, the population problem was essentially eliminated from early or primitive human groups by the same natural factors that control most animal populations, and by the limits placed by man himself through his cultural patterns. *Reproduction became a matter of community interest,* as a major part of the social patterns and adaptations enabling men to live communally. Before the advent of modern technology, environmental factors kept the population limited to areas with a large natural supply of plant and animal foods. Cultural evolution of social restrictions limited population size in a voluntary sense, though the society's rationale for these restrictive factors was often not directly associated with population control *per se*.

Population Growth in Modern Societies

The scientific-industrial revolution marked the beginning of a great upsurge in population growth (Fig. 9-1). As knowledge of medicine and social hygiene spread, more people lived after birth even if in "more misery," as Thomas Malthus cynically said in 1798. Along with this growth, the new revolution brought some degree of knowledge about birth control methods. With the concomitant change in motivation for smaller families engendered by industrialization, the birth rate in technologically advanced countries soon began to decline. To understand what occurred at this time and why, despite the hopes of some demographers, the same transition will not take place in currently underdeveloped major nations, let us review three basic demographic concepts.

1. $\text{Birth Rate} = \dfrac{\text{number of live births}}{\text{1,000 members of population}}$

2. $\text{Death Rate} = \dfrac{\text{number of deaths}}{\text{1,000 members of population}}$

3. $\text{Rate of Natural Increase} = \text{BR} - \text{DR}$

A birth rate of 50 live births annually per 1,000 population is considered very high. In 1971 several countries in Africa (Niger, Rwanda, and Swaziland) had the extraordinary birth rate of 52 per 1,000. A low birth rate today is around 15 per 1,000; in 1971 two European countries, Sweden and Luxembourg, had the world's lowest birth rate of 13.5 per 1,000.

Current death rates range between a maximum average of 30 deaths per 1,000 (Angola in middle Africa) and a minimum of 5 deaths per 1,000 (Singapore, Fiji, Taiwan, and the Ryukyu Islands). On a *worldwide basis,* the annual birth rate in January 1971 was 34 per 1,000 and the annual death rate was 14 per 1,000. This gives us an annual rate of increase of the world's population of 2.0 percent (34−14=20 per 1,000 growth rate, or 2.0 percent).

What does any rate of increase (say, 2 percent) mean in terms of adding more living bodies to the human population? Assuming no change in growth rate, the number of years required to double a population with a particular growth rate is as follows:

Rate of Increase (Percent)	*Doubling Rate of Population (Years)*	*Example of Country or Region with Rate in 1971*
0.1	700	East Germany
0.5	140	Denmark
1.0	70	Japan, United States (both 1.1 percent, 63 years)
1.5	47	Argentina
2.0	35	World, Namibia (Southwest Africa)
2.5	28	Afghanistan, Laos
3.0	24	Ghana, Iran
4.0	17.5	Costa Rica (3.8 percent, 19 years)
8.2	9	Kuwait

Historically, of course, growth rates have fluctuated considerably. Up to the year 1 A.D., the world's population increased at a rate of about 0.002 percent a year. By 1900–1925 the rate was approximately 1.0 percent, and today it is 2.0 percent. The most dramatic upsurge, as we have said, started around 1650 A.D. With our demographic points in mind, let us go back to that time now and look at the peculiar combination of circumstances in Europe that lowered the death rate and then lowered the birth rate, so that about two hundred years after the death rate began to decrease, the overall growth rate became relatively low.

The age-old limiting factors of famine and disease began to be alleviated in seventeenth-century Europe by technological developments in agriculture and food transportation and by medical advances that conquered most of the devastating epidemic and common childhood diseases. Progress in instituting public sanitation measures such as sewer systems and garbage collections reduced mortality factors. Through the next two centuries, these scientific and medical discoveries were exported to other parts of the world through missionaries, traders, and the colonies founded by leading European nations. With the drop in death rate at home and abroad, the rate of population growth rapidly increased.

In the population of western Europe, however, the birth rate began to decline starting shortly after 1800, following in a general way the declining death rate (Fig. 9-2). This diminution of the birth rate paralleling the lowering of the death rate has been called the **demographic transition.** It is thought that this decline in fertility was caused principally by the change in social attitudes towards children. With the progression from an agrarian society to an urban, industrial society, children became less of an economic asset and more of an economic burden. Children were useful workers on a farm, but in the city a father was the principal wage-earner, older children worked for very low wages, and young children prevented a mother from working. Under the crowded urban conditions and the socioeconomic pressures of industrial work, a small family became the most desirable state, and man was motivated to curb his birth rate by whatever means he could.

In the underdeveloped countries, the demographic situation at the mid-twentieth century showed uniformly higher birth rates than in pre-industrial Europe, continuing without any decrease, and a sharply decreasing death rate, the result of the introduction of modern health measures (Fig. 9-3). Instead of being invented and introduced over centuries as in Europe, purification of water supplies, vaccination campaigns, and malaria spraying campaigns were initiated almost

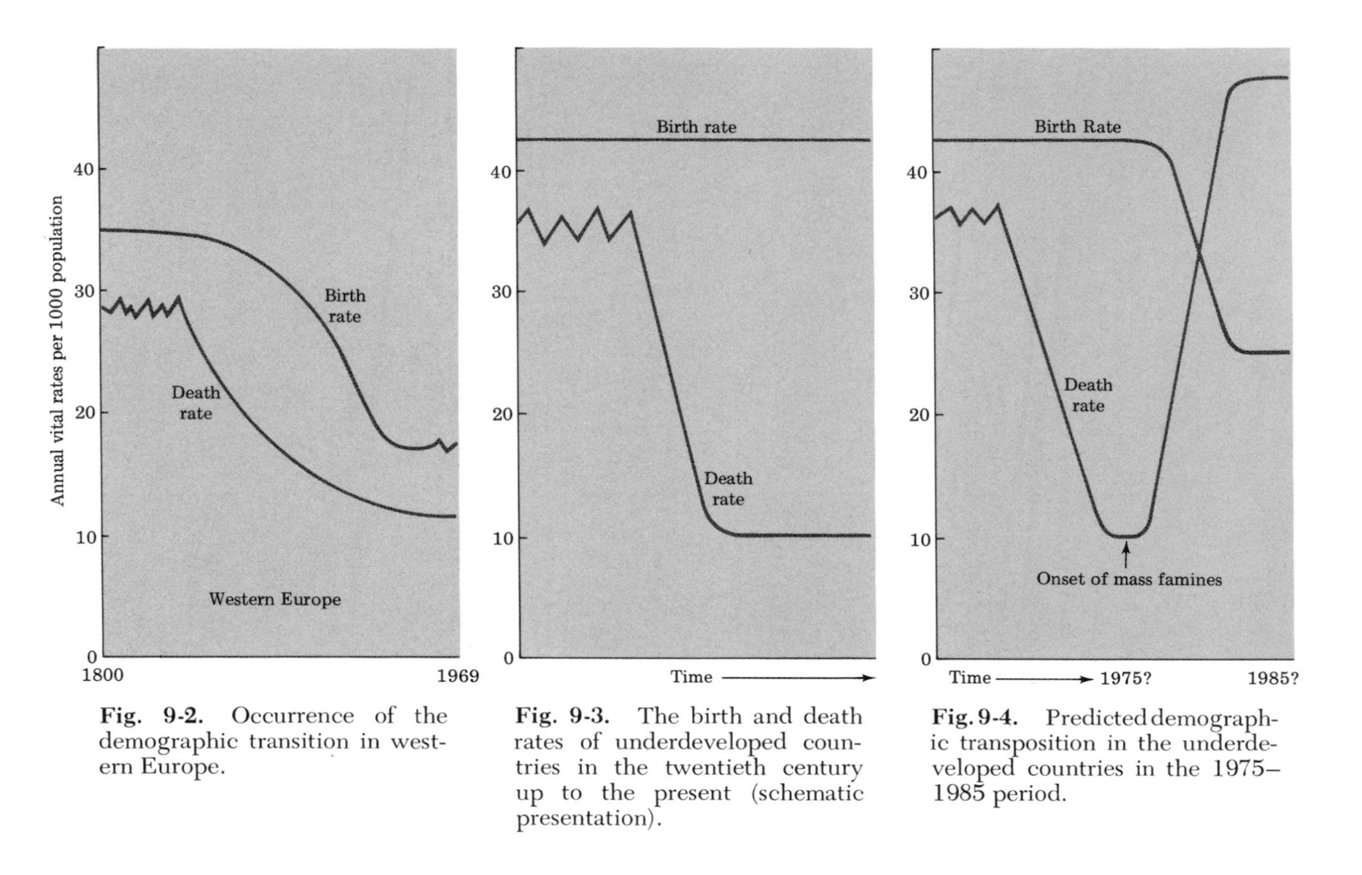

Fig. 9-2. Occurrence of the demographic transition in western Europe.

Fig. 9-3. The birth and death rates of underdeveloped countries in the twentieth century up to the present (schematic presentation).

Fig. 9-4. Predicted demographic transposition in the underdeveloped countries in the 1975–1985 period.

simultaneously in a few years' time. Thus today the demographic situation in these underdeveloped countries has been rapidly brought to a point that superficially resembles western Europe's population parameters around 1700–1800 A.D.

While one might expect to see a continuing parallel in population trends between Europe and underdeveloped countries, sociologists and demographers frequently question whether there *will* indeed be a "demographic transition" in the underdeveloped countries. That is, is it possible that the population problem will solve itself through a *natural* transition of motivation towards smaller families? Many population biologists think this will *not* occur, for the following reasons.

At the time of the demographic transition in Western Europe, the population was small enough so that industrialization and urbanization affected the social values of nearly everyone in those countries. The problem in India and other underdeveloped areas is that their populations are already so large that the majority of the people will simply not be affected by the current level of socioeconomic progress before food crises become severe. Thus biologists such as Paul R. Ehrlich in his book *The Population Bomb* and William and Paul Paddock in their book *Famine—1975!* have predicted that the death rate will greatly increase again within the next decade, without an earlier drop in the birth rate. While scores of millions die, the birth rate among the remaining populace will decrease because of malnutrition and the lack of desire to bring children into a starving society. Michael M. Sligh has suggested in a recent article that this will cause a **demographic transposition** effect in that the relative positions of the death rate and birth rate will be transposed, at least for a time (Fig. 9-4). This demographic transposition will have tragic immediate effects on the populations of many underdeveloped countries, but it may require tragedy on such an immense scale to shock the world into full adoption of population control and zero population growth.

It appears unlikely at this point that worldwide population growth will be curtailed by the same changes in economic and social values that led to the rise of the small family as a norm in Europe. Likewise, it seems that alternatives to the Malthusian parameters of war, famine, and pestilence will have to be sought. War as an overall population control factor will continue to be insignificant unless we have an atomic world war; in that event we would probably be wondering about our very survival as a species. The biological restraints that have always exercised some degree of population control, including famine and epidemic diseases, will undoubtedly continue to operate and, as has been mentioned, massive famines in the near future are predicted by some. Epidemic diseases caused by new mutant viruses may arise and sweep the world or parts of it at any time. In 1969 and early 1970 an extremely virulent virus called Lassa fever was discovered in northern Nigeria and proved so deadly that American researchers had to halt work on it. Fortunately, its spread in Nigeria abated spontaneously in March 1970. Dr. Hans Zinnser, in his book *Rats, Lice and History,* traced the history of epidemic diseases such as smallpox, influenza, and typhus and presented considerable evidence suggesting that infectious viral agents are constantly changing in character and pathogenic effect. In recent years we have become familiar with new flu viruses that sweep across the world and are replaced the next winter by new strains.

From even this brief review it is obvious that some means of voluntarily limiting population size would be far preferable to these disasters. Fortunately, a number of methods for population control are currently available; many of these are promoted as "family planning" methods, which can be used for timing conceptions in marriage. In the following pages we shall look at these products of the scientific-medical revolution and assess their relative merits for worldwide implementation and success as control measures.

Contraception and Family-Planning Programs

Contraceptive devices and methods have been designed to prevent conception by mechanical, chemical, hormonal, or "natural" means. Their economic cost and effectiveness in achieving the desired end vary widely.

Mechanical Contraceptives

These are devices that prevent conception by blocking passage of the sperm into the uterus or by preventing implantation of the fertilized egg in the uterine wall. The **condom** is a very thin latex sheath that fits tightly over the penis and prevents semen from entering the vagina at ejaculation. It requires no prescription and is widely sold (and distributed in the armed forces) as a means of preventing venereal disease. It is relatively inexpensive in the United States, but the cost is still prohibitive for general distribution in most underdeveloped countries. There are aesthetic and psychological disadvantages to its use; yet in the late 1960s it was estimated that 25 percent of all married couples in the United States used this method of contraception.

The **diaphragm** is a rubber cup with a spring steel rim that can be inserted in the vagina before intercourse to cover the cervical entrance to the womb, or uterus. The edges of the diaphragm are coated with a spermicidal agent to prevent sperm from slipping underneath the rim. It is removed several hours after intercourse. Available by prescription in the United States, the diaphragm has to be fitted (at a cost of about $15) by a doctor; the device itself costs about $2. The cost and relatively complicated routine of fitting and periodic insertion prevent its utilization in mass population control programs in the underdeveloped countries. The **cervical cap** is quite similar to the diaphragm and may be worn for three weeks at a time, with removal necessary only near and during the menstrual period. It is extremely effective. The principal disadvantage is the necessity of placing it correctly.

The **intrauterine device** (or **IUD**) is a plastic or metal object that may be inserted into the uterus cavity and left there to prevent conception as long as desired. IUDs are designed in various shapes. They probably act by dislodging the fertilized egg or by preventing its implantation in the uterine wall. It is also suggested that the presence of the foreign object in the uterus hastens the muscular movement of the oviducts and speeds the ovum to the uterus, preventing it from being fertilized en route. The failure rate of the IUD is among the lowest for all contraceptives: about one or two women out of a hundred using an IUD will conceive. The cost is minimal.

Chemical Contraceptives

These include sprays, foams, tablets, suppositories, creams, and jellies designed to kill or immobilize sperm before conception occurs. They are inserted into the vagina before intercourse. They are effective about 60 percent of the time, hence not very reliable for population control.

Hormonal Contraceptives

These contraceptives are the *most effective* in present use. They are female pills or injections to prevent ovulation by simulating the conditions of early pregnancy in the body. The steroid female hormones **estrogen** and **progestin** (a synthetic chemical similar to the natural hormone progesterone, which is produced by the ovaries) are administered either sequentially or in combined form for three weeks of the 28-day menstrual cycle. The pill method is simple, inexpensive, and virtually 100 percent effective. The only requirement is motivation to remember to take the pill once a day. About one woman in every four or five taking the pill experiences undesirable side effects. Male steroid hormonal pills are as yet unsatisfactory because of side effects.

"Natural" Contraceptive Methods

The **rhythm method,** or periodic abstention, is the only method of birth control approved by the Roman Catholic Church. Since the timing of ovulation (release of the mature egg by the ovary into the oviduct, where it can be fertilized) is usually 16 days before the menstrual period, one can abstain from intercourse from several days before ovulation until about a day after ovulation and avoid fertilization of the egg. Unfortunately, the timing of ovulation varies among women and even from month to month in the same woman; in fact, one out of six women has such an irregular cycle that the method will not work at all for her. The rhythm method has a notoriously high failure rate, both because of this irregularity of the cycle and because of the occasional lapse of discipline on the part of the participants.

Other "natural" contraceptive methods include **coitus interruptus, douching,** and **abstention** from intercourse altogether.

Overall, most of these contraceptive methods have been promoted on the premise that "family planning" was the goal to be reached. That is, using one or more of these procedures would allow a couple to decide how many children they would have rather than trusting to chance alone. So despite the high effectiveness of the mechanical, chemical, and hormonal devices, the use of contraceptives has proven successful to date solely as a population growth *depressant,* not a population *control.* The basic problem is to achieve an international *motivation* for small families of at most two children. There is no doubt that effective contraceptive *methods* are already available to stop population growth.

Abortion

If contraceptive methods fail and a woman conceives, the commonest method employed around the world to prevent the birth of an unwanted child is induced or therapeutic **abortion** (expulsion of the fetus from the uterine cavity). Up to about 12 weeks of fetal development, this can be done without much chance of injury to the mother. After about 16 weeks, however, the risks of death to the mother by hemorrhage become considerable. The medically approved method of abortion is dilation (of the cervix) and curettage (scraping) of the uterus; a physician working under sterile conditions can do a "D and C" with an early fetus with less risk than delivering a full-term pregnancy. Cruder forms of abortion are practiced in every nation in the world, and abortion represents the single greatest factor contributing to the death rate among pregnant women. In fact, one reason why abortion was made illegal in the nineteenth century was to protect the mother from its danger under the primitive medical conditions of that time.

In recent years there has been an international trend toward legalizing abortions, based on the premise that the decision whether to have a child should be left up to the individual, not society. The opponents of this premise state that the third entity involved, the unborn child, should be protected by society's laws from the unilateral act of abortion approved by the mother. While lawyers and religious groups differ on defining the point in development at which a fetus becomes a human person complete with a soul, opponents of abortion feel that it denies the child's right to be born. His parents had the right to decide whether to conceive him, but once conceived, his right to live should not be prejudiced. Obviously, the best solution lies in effective contraception, which avoids the supposed necessity of abortion.

Sterilization

Sterilization of the man or woman is 100 percent effective as a birth-control method, and has no side effects on sex life or normal physiology. Naturally, most sterilizations are obtained by couples who have completed their families and wish to avoid the complications of contraceptives. The operation in the male involves tying off or cutting the vas deferens tubes leading from the testes to the uretha tube in the penis; thus the ejaculate no longer contains sperm cells. This **vasectomy** operation through the scrotal sac wall takes about fifteen minutes and can be done very inexpensively under local anesthetic in a physician's office or field clinic.

An analogous operation, called a **salpingectomy,** can be done in the female. However, this involves special abdominal surgery under anesthesia, unless the operation on the fallopian tubes is done when they are accessible following the birth of a baby. In either the sterilized male or female, sperm and eggs continue to be produced regularly but eventually break down and are reabsorbed by the body or picked up by phagocytes (specialized white blood cells) from the blood. There is at least a 50 percent chance that either a vasectomy or a salpingectomy will be reversible.

In a third operation resulting in sterilization, a **hysterectomy,** the uterus itself is removed. This is rarely done for birth-control purposes alone but is a rather common operation in women experiencing menopause.

More severe sterilization operations include **castration** (removal of the testes) of the male, and **oöphorectomy** (removal of the ovaries) in the female. These operations result in detrimental hormone imbalance because the sex organs are endocrine glands producing hormones that control the secondary sexual characteristics of an individual. Without these hormones, the body may experience serious alterations. Vasectomy and salpingectomy, of course, only prevent fertilization; they do *not* prevent hormonal production or upset any other normal bodily activity.

Male sterilization has proven a useful part of population control programs in parts of India and other countries, but western cultures have so far not readily accepted this simple contraceptive measure. This is largely the result of ignorance of the exact nature of the male operation and fears that a man's sex life will end. Obviously, these fears are completely unfounded. Vasectomies would seem to be an effective and easy population-control measure to advocate among married men who already have one or two children and do not wish a larger family.

Prospects for the Future

The present impact of man's population growth on natural communities, on the continued existence of other species, and on his own biological and psychological character, all shout in the loudest possible way the dangers of continued increase. Imaginative proposals have been made to feed and house the additional billions of humans soon to join our "standing crop" of Man. These include using artificial symbioses between nitrogen-fixing bacteria and non-legume plants to utilize more directly the vast reservoir of molecular nitrogen in the atmosphere, or increasing the production of food from the sea, or perhaps artificially synthesizing protein foods from algal cultures on a massive scale. Trans-planetary migration of humans via spaceship has even been suggested to ease future crowding on earth. Yet objective analyses of all these ideas show that they are but proximate solutions that can only handle a minute fraction of our estimated future increases in population. The ultimate solution to the crises we face between man and his earthly environment is, bluntly, *zero* population growth. Man has the choice of reaching agreement on the desirability of this goal and implementation of strategy to reach it, or continuing as he is with the consequences becoming now all too familiar to us.

Summary

Population growth in man has been caused primarily by a cultural revolution (600,000 B.C.), the agricultural revolution (6000 B.C.), and the scientific-industrial revolution that started about 1650 A.D. The last has caused the "population explosion" because of the sizable gap created between our birth rates and death rates around the world. Population growth in primitive hunting-and-gathering societies was prevented or minimized by famine, disease, war, and many culturally instituted behavioral patterns (e.g., infanticide and abortion). In modern societies medical advances and social hygiene lowered the death rate while the birth rate remained generally high. For the world in January 1971, the annual birth rate was 34 per 1,000 and the death rate 14 per 1,000, giving us an annual rate of increase of 2.0 percent, which will double the world's population every 35 years. It is doubtful that the demographic transition that reduced growth rates in European countries will take place in presently underdeveloped countries with large populations. Instead, a demographic transposition may appear in these countries because of deaths on a massive scale. True population control is possible if the people of the world develop the *motivation* to have small families (average of 2 children). The technological measures for voluntary achievement of the goal of zero population growth are available in the form of contraceptives, abortion, and sterilization techniques.

Additional Reading

The number of books and articles being published each year on environmental problems staggers the imagination *and* the bookshelves of libraries. While it may seem presumptuous to select a small list of these for suggested additional reading, your interest may have been aroused sufficiently so that you would like to pursue in more depth some of the topics we have touched upon in this book. The following offerings are grouped according to chapter.

1. POPULATIONS AND ECOLOGY (General Books)

Boughey, Arthur S. 1971. *Man and the Environment.* Macmillan, New York. 472 pp.

Ehrlich, Paul R., and Anne H. Ehrlich. 1972. *Population, Resources, Environment.* Second edition. W. H. Freeman, San Francisco. 509 pp.

Wagner, Richard H. 1971. *Environment and Man.* W. W. Norton, New York. 491 pp.

2 and 3. SOME BASIC PRINCIPLES OF ECOLOGY

Allee, W. C.., Alfred E. Emerson, Orlando Park, Thomas Park, and Karl P. Schmidt. 1949. *Principles of Animal Ecology.* W. B. Saunders, Philadelphia. 837 pp.

Kendeigh, S. Charles. 1961. *Animal Ecology.* Prentice-Hall, Englewood Cliffs. 468 pp.

Odum, Eugene P. 1971. *Fundamentals of Ecology.* Third edition. W. B. Saunders, Philadelphia. 574 pp.

Smith, Robert L. 1966. *Ecology and Field Biology.* Harper & Row, New York. 686 pp.

4. THE ORGANIZATION OF POPULATIONS

Andrewartha, H. G., and L. C. Birch. 1954. *The Distribution and Abundance of Animals.* University of Chicago Press, Chicago. 782 pp.

5. POPULATION GROWTH AND REGULATION

Slobodkin, Lawrence B. 1961. *Growth and Regulation of Animal Populations.* Holt, Rinehart and Winston, New York. 184 pp.

6. THE ECOLOGY OF COMMUNITIES

Shelford, Victor E. 1963. *The Ecology of North America.* University of Illinois Press, Urbana. 610 pp.

Whittaker, Robert H. 1970. *Communities and Ecosystems.* Macmillan, New York. 162 pp.

7. ENVIRONMENTAL ALTERATION

Ehrenfeld, David W. 1970. *Biological Conservation.* Holt, Rinehart and Winston, New York. 226 pp.

Owen, Oliver S. 1971. *Natural Resource Conservation: An Ecological Approach.* Macmillan, New York. 593 pp.

Weisberg, Barry. 1971. *Ecocide in Indochina: The Ecology of War.* Canfield Press, Harper & Row, San Francisco. 241 pp.

8. POLLUTION

Cox, George W., editor. 1969. *Readings in Conservation Ecology.* Appleton-Century-Crofts, New York. 595 pp.

Detwyler, Thomas R., editor. 1971. *Man's Impact on Environment.* McGraw-Hill, New York. 731 pp.

9. POPULATION GROWTH IN MAN

Ehrlich, Paul R. 1969. *The Population Bomb.* Sierra Club Books, San Francisco. 191 pp.

Emmel, Thomas C., and Michael M. Sligh. 1970. Human population problems. *Science Education,* 54:363–372.

Malthus, Thomas Robert. 1798. *On Population.* Reprinted edition (1960) by Modern Library, New York. 602 pp.

Zinsser, Hans. 1935. *Rats, Lice and History.* Bantam Books, New York. 228 pp.

Glossary

The definitions of terms in this glossary pertain specifically to the use of these words in ecology and population biology. Hence some terms that have other meanings outside these fields will be found to have rather restricted definitions here.

Abortion—expulsion of a fetus from the uterine cavity before normal birth.

Age Distribution—proportion of a population in each age class.

Aggressive Mimicry—a situation where one species mimics another species or an object in its environment in order to attract or deceive the prey species for the purpose of eating it.

Allopatric—occurring in different areas.

Arboreal—living in trees.

Autecology—the study of the ecology of the individual organism of a species.

Batesian Mimicry—a mimicry situation in which a harmless species (mimic) resembles a harmful species (model) and thus gains protection from predation.

Biogeochemical Cycle—the cyclical series of transformations of a chemical element through the organisms in a biotic community and their physical environment.

Biological Control—the suppression of reproduction of a pest organism by utilizing behavioral traits or other organisms rather than chemical means.

Biomass—the weight of all living organisms in a sample.

Biome—a major climax community type covering a wide area of a continent or the earth.

Biosphere—all living organisms on earth.

Biota—the flora and fauna of a region.

Biotic—relating to life; biological.

Biotic Potential—maximum possible growth rate of living things under ideal conditions.

Bulkhead Line—a legally set limit on commercial filling operations in a bay.

Carrying Capacity—the maximum population size that the environment can support without deterioration.

Carnivore—an animal that feeds on other animals.

Chaparral—a biome consisting of dense thickets of stiff or tough-leafed shrubs, found in areas with a Mediterranean climate.

Character Displacement—an outcome of competition in which two species living in the same area have evolved differences in morphology or other characteristics that lessen competition for food resources.

Clear-cutting—logging all trees in a forest area.

Climax Community—the terminal stage of ecological succession in an area.

Commensalism—a symbiotic relationship in which one species benefits and the other is neither benefited nor harmed.

Community—all the organisms of all species living in a particular area.

Competition—a struggle between individuals of the same or different species for food, space, mates, or any other limited resource.

Competitive Exclusion—a result of competition where one species is forced out of part of the available habitat by a more efficient species.

Contraception—the process of preventing conception (successful union of a sperm and egg cell).

Cryptic Coloration—coloration that causes an organism to resemble some inanimate object or background substrate.

Decomposers—organisms such as carrion beetles and fungi that feed upon and break down dead organic matter.

Demographic Transition—a decline in birth rate following a drop in death rate, as in the case of the population of Western Europe from 1650 to 1850.

Demographic Transposition—a predicted reversal of the present relative positions of the death and birth rates in major underdeveloped countries around 1975–85; or, any similar population trend when a country experiencing population growth reaches the point of being unable to supply the subsistence-level needs of a significant proportion of the population.

Density—the size of the population within a particular unit of space.

Desert—an arid biome characterized by little moisture, high diurnal temperatures, and low diversity of vegetation and animal life.

Dispersal—spreading or moving out from a point.

Dispersion—the internal distribution pattern of individuals in a population.

Disseminules—seeds, spores, eggs, or other non-adult stages of plants and animals that can be dispersed.

Dominance—the fact of being a major influence on numerical composition or internal energy dynamics in a community.

Dominant Species—a species of plant or animal that is particularly abundant or controls a major portion of the energy flow in a community.

Ecological Efficiency—the percent of available energy utilized by a trophic level from the next lowest level.

Ecological Pyramid—a triangular graphic form showing the sequence of diminishing numbers, biomass, or available energy as one goes up through the trophic levels.

Ecology—the study of the interrelationships between organisms and their environment.

Ecosystem—the biotic community and physical environment in an area.

Emigration—departure from one place to take up residence in another area.

Environment—the physical and biological characteristics of a certain region.

Environmental Resistance—all the limiting factors in the environment that are acting on a particular population.

Epiphylls—tiny plants such as mosses that grow on the surfaces of leaves of larger plants in tropical rain forests.

Epiphyte—a non-parasitic plant that grows upon other plants for physical support, deriving its required moisture chiefly from the air. Epiphytic plants include arboreal orchids, bromeliads, and many mosses and ferns.

Eutrophication—promotion of plant growth in an aquatic ecosystem by rapidly adding substantial amounts of nutrients. Available oxygen is used up by the plants and substantial fish kills may result.

Evolution—a change in gene frequency in a population, usually involving a visible change in the species' characteristics.

Extinction—an outcome of competition and natural selection in which the species or population dies out completely.

Food Chain—the transfer of food energy from green plants (ultimately from the sun) through a series of species.

Food Web—the complex interlocking patterns of food chains in a biotic community.

Gause's Principle—two species cannot occupy the same niche simultaneously.

Grassland—a treeless, grassy plains biome found in central parts of temperate North America, Eurasia, and Australia.

Growth Form or Curve—the manner and speed of population increase.

Habitat—the physical place where an organism lives.

Herbivores—animals that eat plants.

Home Range—physical area of an individual's normal activity.

Host—an organism that is fed upon by another organism.

Immigration—one-way movement into a new area of residence from neighboring areas.

Infanticide—killing a human infant after birth.

Laterite—a hard, impermeable, reddish crust of soil formed by water leaching exposed soils in cleared areas of tropical rain forests.

Leaching—dissolving out of soluble materials (such as salts and minerals) by water percolating through soil, especially in areas of heavy rainfall.

Migration—periodic departure and return of individuals to and from a population area.

Morphology—the form and structure of animals and plants.

Mortality—death rate.

Mullerian Mimicry—a situation in which all potential prey species in a mimicry complex are harmful and share a common warning color pattern and behavior.

Mutualism—a symbiotic relationship where both species benefit.

Natality—birth rate.

Niche—the unique occupation or way of life of a plant or animal species: where it lives and what it does in the community.

Nitrogen-fixing Bacteria—bacteria with the ability to convert inorganic N_2 into forms, especially nitrates, that are immediately usable by plants.

Oscillations—periodic fluctuations in population size.

Parasite—an organism that feeds on another organism (its host), generally without killing it.

Parasitism—a symbiotic relationship where one species (the host) is harmed but not killed immediately and the other species feeding on it (the parasite) is benefited.

Periodism—a particular temporal sequence of activity.

Pesticide—a chemical used to kill species regarded as pests.

Pollution—the introduction of an undesirable change in some aspect of the environment.

Population—an aggregation of organisms of the same species living in the same area.

Primary Production—the energy accumulated and stored by plants through photosynthesis.

Predator—an organism that kills and eats another organism.

Producers—organisms that convert light energy from the sun into chemical-bond energy.

Productivity—the production of organic matter per unit of area in a community over a particular time interval, usually a year.

Saprophyte—a plant, such as a fungus or mushroom, that feeds on dead and decaying matter.

Savanna—tropical grassland biome with scattered trees.

Sere—the complex series of changes in an area, from the initial condition to the climax community.

Social Hierarchy—a series of dominance-subordination relationships in a group (= "pecking order").

Standing Crop—the number of individuals or total biomass present in a community at one particular time.

Stratification—vertical layering of a plant or aquatic community.

Substrate—the substance or base on which an organism grows.

Succession—the replacement of one community by another.

Survivorship—the percentage of individuals in a population still living at various times after birth.

Symbiosis—the living together in intimate association of two diverse types of organisms.

Sympatric—occurring in the same area.

Taiga—the northern coniferous forest biome, found south of the Arctic Circle.

Temperate Deciduous Forest—the broadleaved forest biome north and south of the equatorial tropics.

Territoriality—the behavior of defending an area against intrusion by other individuals.

Trophic Level—a feeding level. Organisms having the same general source of energy or nutrition are said to be in the same trophic level.

Tropical Deciduous Forest—a tropical biome with well-developed broadleaved forest and seasonal loss of leaves.

Tropical Rain Forest—the most complex biome on earth, characterized by great species diversity, high rainfall all year, and warm temperatures. The broadleaved forest remains green throughout the year.

Tundra—the treeless plains biome of northern arctic regions and high alpine mountaintops.

Turbidity—cloudiness of water, from the presence of suspended sediment or small organisms such as algae.

Vagility—a species' inherent ability to move.

Warning Coloration—bright coloration advertising the fact that a potential prey species is harmful (distasteful, poisonous, irritating, etc.).

INDEX

Explanations of terms that appear in figure captions or are illustrated by photographs are identified by *italicized* numbers.

Emmel, Thomas C

An introduction to ecology and population biology [by] Thomas C. Emmel. [1st ed.] New York, Norton [1973]

x, 196 p. illus. 22 cm.

Bibliograpy: p. 176-177.

1. Ecology. 2. Population. 3. Man—Influence on nature. I. Title.

QH541.E45 1973 574.5′24 72-14170

ISBN 0-393-06393-3 : 0-393-09371-9 (pbk.) MARC

Library of Congress 73 [4]